JUVENILE JUSTICE AND DELINQUENCY

Sara Miller McCune founded SAGE Publishing in 1965 to support the dissemination of usable knowledge and educate a global community. SAGE publishes more than 1000 journals and over 800 new books each year, spanning a wide range of subject areas. Our growing selection of library products includes archives, data, case studies and video. SAGE remains majority owned by our founder and after her lifetime will become owned by a charitable trust that secures the company's continued independence.

Los Angeles | London | New Delhi | Singapore | Washington DC | Melbourne

JUVENILE JUSTICE AND DELINQUENCY

Barry A. Krisberg

University of California, Berkeley

Los Angeles | London | New Delhi
Singapore | Washington DC | Melbourne

FOR INFORMATION:

Sage Publications, Inc.
2455 Teller Road
Thousand Oaks, California 91320
E-mail: order@sagepub.com

Sage Publications Ltd.
1 Oliver's Yard
55 City Road
London EC1Y 1SP
United Kingdom

Sage Publications India Pvt. Ltd.
B-42, Panchsheel Enclave
Post Box 4109
New Delhi 110 017 India

ISBN: 978-1-5063-2923-9

Acquisitions Editor: Jessica Miller
eLearning Editor: Laura Kirkhuff
Production Editor: Libby Larson
Copy Editor: Diane Wainwright
Typesetter: Hurix Systems Pvt. Ltd.
Proofreader: Theresa Kay
Indexer: Molly Hall
Cover Designer: Gail Buschman
Marketing Manager: Amy Lammers

Printed on acid-free paper in the United States of America.

17 18 19   20 21 10 9 8 7 6 5 4 3 2 1

CONTENTS

PREFACE

More than 30 years ago, I had the unique opportunity of spending almost every day for several months with about 20 young men who were leaders of juvenile gangs in Philadelphia. They introduced me to their fellow gang members and family members and allowed me to share part of their world. They generously shared their lives with me and profoundly changed my own.

To law enforcement officials and to many community residents, these young men were very dangerous offenders who had been arrested many times and had spent a fair amount of time in juvenile facilities and adult prisons, but to me, they were my teachers and mentors in settings that were very different from my experiences. First and foremost, these young men taught me to listen and to reflect on the life experiences of those less fortunate than myself. Much of my academic and professional career has been built on listening carefully to young people who are often very angry and bitter about their own lives. I have always tried to faithfully communicate what they told me to others in the worlds of public policy, the university community, professional groups, and students. Most recently, I found myself interviewing more than 150 young people who were confined in the California Youth Authority. They told me of everyday examples of abuse and maltreatment by those who were supposed to provide for their education and treatment. Retelling their stories to my fellow Californians is leading to a major reexamination of how the nation's most populous state is treating its troubled young people. I saw the reformative power of this approach in the early 1970s when Jerome Miller, then Commissioner of the Massachusetts Department of Youth Services, held press conferences across the state and simply asked his young clients to tell the media what was being done to them by a brutal and corrupt youth corrections system. One result of this effort was that Massachusetts closed its barbaric training schools and returned most youths to smaller, high-quality, community-based programs. Even today, the Miller reforms represent the "gold standard" for juvenile justice in the Bay State and across the nation.

My Philadelphia encounters with young gang members taught me to embrace the value of redemptive justice. It is what we all would want for ourselves and our family members—a justice system that offers the hope that people can improve and can restart their lives in a more positive direction. I believe that the concept of redemptive justice was central to what Jane Addams and other pioneers in juvenile justice were striving to achieve. Rather than killing off the idealistic vision of the juvenile court, we need to rediscover it.

This book attempts to assemble the research that supports this perspective. I hope to encourage students to think critically. Sometimes, the language used in this book is less than polite—but now is the time for plain talk and "speaking truth to power." My hope is that this book and the courses in which it will be utilized will be springboards to lead the current generation of idealistic young people into progressive action.

I have many debts to repay to those who helped me assemble this book. National Council on Crime and Delinquency researchers Priscilla Aguirre, Jessica Craine, Sharan Dhanoa, Poonam Juneja, Kelly Knight, and Susan Marchionna assisted in reviewing earlier drafts, running down key references, and suggesting ways to improve the overall effort. My appreciation goes to great colleagues such as Yitzhak Bakal, Frederick Mills, Richard Tillson, and Buddy Howell who were my sounding boards, patiently listening as I tried to work out my ideas. I owe a special debt to the University of California at Berkeley students who lived through classes devoted to the materials in this book and showed me how to improve my presentation. I also wish to express my appreciation for academic colleagues at the Law School and the Institute for the Study of Social Issues.

To my own teachers and mentors, especially Marvin Wolfgang and Thorsten Sellin, I owe so much. They taught me that ideas could change the world for the better and that advances in enlightened social policy are the true measure of lasting intellectual contributions.

Jerry Westby of Sage Publications was a patient and very supportive editor. Susan Marchionna was the central person in pulling together this manuscript. She conducted complex research, forced me to write and think more clearly, and made sure that the final product was worthy of its goals. Without Susan's very hard work, this book would still be in progress.

Finally, I want to thank my sons, Moshe and Zaid, and my daughter-in-law Jessica, who continue to keep me plugged into the world of young people, educating me about what is really important in life. My life partner, Karen McKie, is responsible for infusing a humanistic spirit into my scholarly work. Her "conceptual physics" helped me look at familiar topics with a fresh set of eyes. Most important, she has showed me what unconditional love and building community can accomplish in a less-than-perfect world. Her love and support make my world possible.

Sage Publications gratefully acknowledges the contributions of the following reviewers of earlier editions of the book:

Joanitha Barnes,
Southwest Tennessee Community College

Ed Bowman,
Lock Haven University of Pennsylvania

Margaret Pate,
Radford University

Judith A. Ryder,
St. John's University

Michelle C. Watkins,
El Paso Community College

Franklin E. Zimring,
University of California at Berkeley

Andrea L. Kordzek,
University at Albany

Yolander G. Hurst,
Southern Illinois University

Charles E. Owens,
University of North Florida

Patricia H. Jenkins,
Temple University

Nancy Rodriguez,
Arizona State University, West

Sesha Kethineni,
Illinois State University

Morgan Peterson,
Palomar College

Lee Ayers,
Southern Oregon University

Charles L. Dreveskracht,
Northeastern State University

Verna J. Henson,
Texas State University–San Marcos

Martha Smithey,
University of Texas at El Paso

Allison Ann Payne,
College of New Jersey

Frances G. Pestello,
University of Dayton

Ron Fagan,
Pepperdine University

Jeffrey P. Rush,
University of Tennessee at Chattanooga

Terrance J. Taylor,
Georgia State University

ACKNOWLEDGMENTS

I would like to thank my colleagues at the Law School and the Institute for the Study of Societal Issues at the University of California, Berkeley, who provided me with encouragement and assistance in the writing of this book. Many of these colleagues indulged my early ideas about how to approach this project.

Much praise goes to Jerry Westby and Laura Kirkhoff at Sage Publications who shepherded this book from its conception to the final manuscript. I am also indebted to Professor Carly Dierkhising who had valuable suggestions about improvements that could be made in this edition.

Most of all, I must thank my children, Moshe, Zaid, and Jessica, whose love, support, and fresh perspectives nurture me. To my life partner, Karen McKie, who taught me about the power of unconditional love and compassion, I owe so much.

This book, which is very much about the future, is dedicated to my granddaughter, Memphis McKie-Krisberg, who reminded me of the power of discovery, joy, and unyielding spirit to change the world.

Chapter 1

JUVENILE JUSTICE

Myths and Realities

"It's only me." These were the tragic words spoken by Charles "Andy" Williams as the San Diego Sheriff's Department SWAT team closed in on the frail high school sophomore who had just turned 15 years old. Williams had just shot a number of his classmates at Santana High School, killing two and wounding 13. This was another in a series of school shootings that shocked the nation; however, the young Mr. Williams did not fit the stereotype of the "superpredator" that has had an undue influence on juvenile justice policy for decades. There have been other very high-profile cases involving children and teens that have generated a vigorous international debate on needed changes in the system of justice as applied to young people.

In Birmingham, Alabama, an 8-year-old boy was charged with "viciously" attacking a toddler, Kelci Lewis, and murdering her (Binder, 2015). The law enforcement officials announced their intent to prosecute the boy as an adult. The accused perpetrator would be among the youngest criminal court victims in U.S. history. The 8-year-old became angry and violent, and beat the toddler because she would not stop crying. Kelci suffered severe head trauma and injuries to major internal organs. The victim's mother, Katerra Lewis, left the two children alone so that she could attend a local nightclub. There were six other children under the age of 8 also left alone in the house. Within days, the mother was arrested and charged with manslaughter and released on a $15,000 bond after being in **custody** for less than 90 minutes. The 8-year-old was held by the Alabama Department of Human Services pending his **adjudication**.

A very disturbing video showed a Richland County, South Carolina, deputy sheriff grab a 16-year-old African American teen by her hair, flipping her out her chair and tossing her across the classroom. The officer wrapped his forearm around her neck and then handcuffed her. It is alleged that the teen refused to surrender her phone to the deputy. She received multiple injuries from the encounter. The classroom teacher and a vice principal said that they believed the police response was "appropriate." The deputy was suspended and subsequently fired after the Richland County Sheriff reviewed the video. There is a civil suit against the school district and the sheriff's department for the injuries that were sustained (Strehike, 2015).

One of the highest profile cases involving juvenile offenders was known as the New York Central Park jogger case (Burns, 2011; Gray, 2013). In 1989 a young female investment banker was raped, attacked, and left in a coma. The horrendous crime captured worldwide attention. Initially, 11 young people were arrested and five confessed to the crimes. These five juvenile males, four African American and one Latino, were convicted for a range of crimes including assault, robbery, rape, and attempted murder. There were two separate **jury** trials, and the defendants were sentenced to between 5 and 15 years. In today's more punitive environment, the sentences would be even stiffer. Appeals were filed, but the original **convictions** were upheld. Then, shockingly, in 2002, another youth, Matias Ryes, confessed that he had actually committed the rape and his claim was verified by DNA evidence. Reyes claimed that he had engaged in the rape and assault by himself and was not part of the five juveniles who had already been imprisoned. At the time of his confession, Reyes was serving a life sentence for several rapes and murder. At the time of the assault, the media and many politicians used the case to frighten the public about "wilding," allegedly involving **gangs** of young people who would viciously attack strangers for no apparent reason. The police violated all sorts of basic rules governing the handling of these juveniles. The names of the arrested youth were given to the media, and their names and photographs were published in local newspapers. The five youth were interviewed and videotaped, some without access to their parents or guardians. Within weeks, all the defendants retracted their confessions, claiming that they had been intimidated and coerced into their admissions. Police told the youth that their fingerprints were found at the crime scene, but as noted earlier, the DNA evidence collected at the crime scene did not match any of the Central Park youth.

The New York district attorney subsequently withdrew the charges against all of the Central Park Five but never declared that they were innocent. The youth had already served between 6 and 13 years in state **prison**. The convicted Central Park men filed a lawsuit against New York City for malicious prosecution and emotional distress. Mayor Michael Bloomberg refused to settle the case over the next 10 years, but the new mayor, Bill de Blasio, supported a settlement and the courts approved a claim for $41 million in damages in 2014. The Central Park men are continuing to litigate against New York State for additional damages. A noted documentary film-maker made a dramatic film about the Central Park Five, but the New York Police Department and the Manhattan district attorney fought hard to discredit the Ken Burns project and to prevent its airing by PBS.

The Central Park Five case illustrates how misinformation and outright falsehoods have played a significant role in the evolving nature of the justice system's response to alleged young offenders. Juvenile justice policies have historically been built on a foundation of myths. From the "dangerous classes" of the 19th century to the superpredators of the late 20th century, government responses to juvenile crime have been dominated by fear of the young, anxiety about immigrants or racial minorities, and hatred of the poor (Platt, 1968; Wolfgang, Thornberry, & Figlio, 1987). Politicians have too often exploited these mythologies to garner electoral support or to push through funding for their pet projects. The general public has bought into these myths, as evidenced by numerous opinion polls illustrating the perception that juvenile crime rates are raging out of control (Dorfman & Schiraldi, 2001). Even during periods in which juvenile arrests were falling, the National Victimization Survey in 1998 reported that 62% of Americans felt that juvenile crime was rising. A 1996 California poll showed that 60% of the public believed that youths are responsible for most violent crime, although youngsters under age 18 years account for just 13% of arrests for violent offenses. Similarly, the public perceives that school-based violence is far more common than the rates reflected in official statistics.

Several observers feel that these misperceptions are, in part, created by distorted media coverage of juvenile crime (Dorfman & Schiraldi, 2001).

By far, the most destructive myth about juvenile crime was the creation of the superpredator myth (Elikann, 1999). The myth began with predictions of future increases in youth violence made by James Q. Wilson (1995) and John DiIulio (1995a). Wilson claimed that by 2010 there would be 30,000 more juvenile "muggers, killers, and thieves." DiIulio predicted that the new wave of youth criminals would be upon us by 2000. Within a year, DiIulio's (1996) estimate for the growth in violent juveniles had escalated to 270,000 by 2010 (compared to 1990). Other criminologists such as Alfred Blumstein (1996) and James Fox (1996) suggested that the rise in violent arrests of juveniles in the early 1990s would combine with a growing youth population to produce an extended crime epidemic. Fox warned that our nation faces a future juvenile violence that may make today's epidemic pale in comparison (Fox, 1996). He urged urgent action. Not to be outdone in rhetoric, DiIulio referred to a "Crime Bomb" and painted the future horror that "fatherless, Godless, and jobless" juvenile "superpredators" would be "flooding the nation's streets" (DiIulio, 1996, p. 25).

All of these dire predictions proved inaccurate. Juvenile crime rates began a steady decline beginning in 1994, reaching low levels not seen since the late 1970s. In part, the myth was based on a misinterpretation of the research of Wolfgang, Figlio, and Sellin (1972), which found that a small number of juveniles accounted for a large number of juvenile arrests. DiIulio and his panicky friends applied this number to the entire growth in the youth population to manufacture their bogus trends. But even worse, the academic purveyors of the superpredator myth used overheated rhetoric to scare the public.

Consider that the definition of a predator is an animal that eats other animals. Perhaps only the *Tyrannosaurus rex* might truly qualify as a superpredator. The symbolism of the vicious youth criminal who preys on his victims is truly frightening. This is reminiscent of the Nazi propaganda that referred to Jews as vermin that spread disease and plague. Further, the imagery of the child without a conscience was reinforced by media accounts of a generation of babies born addicted to crack cocaine and afflicted with severe neurological problems. Interestingly, a recent review of medical studies of "crack babies" found no substantial evidence that in utero exposure to cocaine negatively affected the child's development more than traditional risk factors such as parental alcohol and tobacco consumption. Herrnstein and Murray (1994) completed the grotesque portrait of the superpredator by claiming to demonstrate a linkage between low

CASE STUDY: JUVENILE JUSTICE: MYTHS AND REALITIES

A 17-month-old girl was found dead in her crib from blunt force trauma and internal injuries. An 8-year-old boy was charged by the police as the murderer. The boy had been left to take care of the baby as the infant's mother went out dancing with her friend. There were four other very young children in the house ages 2, 4, 6, and 7. There were no adults in the house when the crime occurred. The 8-year-old admitted that he shook the infant and threw her against the crib to stop her from crying. He was taken into custody by the local child welfare agency pending the disposition of the manslaughter charge. The boy was charged with manslaughter, and his case will be heard in the family court. The infant's mother was arrested and charged with child endangerment, but the charges were soon dismissed.

intelligence and crime. They suggested that persons of low IQ would respond only to blunt punishments rather than more subtle **prevention** or **rehabilitation** programs.

The media loved the dramatic story about the "barbarians at the gates," and the politicians soon jumped on the bandwagon. A major piece of federal juvenile crime legislation enacted in 1997 was titled The Violent and Repeat Juvenile Offender Act of 1997 (S. 10).

At the state level, the superpredator myth played an important role in 47 states amending their laws on juvenile crime to get tougher on youthful criminals (Torbet et al., 1996). Legislators modified their state laws to permit younger children to be tried in adult criminal courts. More authority was given to **prosecutors** to file juvenile cases in adult courts. Judges were permitted to use "blended sentences" that subjected minors to a mixture of juvenile court and criminal court sanctions. Legislators also weakened protection of the confidentiality of minors tried in juvenile courts, allowing some juvenile court convictions to be counted later in adult proceedings to enhance penalties. State laws were amended to add punishment as an explicit objective of the juvenile court system and to give victims a more defined role in juvenile court hearings. Prior to these revisions, victims of juvenile crime had no formal participation in juvenile court proceedings. During the 1990s, rates of juvenile incarceration increased, and more minors were sentenced to adult prisons and jails. This social and legal policy shift and its consequences are discussed further later in this chapter.

The movement to treat ever younger offenders as adults was aided by other myths about juvenile justice. First, it was asserted that the juvenile court was too lenient and that it could not appropriately sanction serious and violent youthful offenders. Second, it was argued that traditional juvenile court sanctions were ineffective and that treatment did not work for serious and chronic juvenile offenders. Neither of these myths is supported by empirical evidence.

An analysis of juvenile court data in 10 states found that juvenile courts responded severely to minors charged with homicide, robbery, violent sex crimes, and aggravated assaults (Butts & Connors-Beatty, 1993). This study found that juvenile courts sustained petitions (the juvenile court equivalent of a conviction) in 53% of homicide cases, 57% of robbery cases, 44% of serious assaults, and 55% of violent sex crimes. By contrast, a study by the Bureau of Justice Statistics of adult felony cases in state courts found that the odds of an arrested adult being convicted for violent offenses ranged from a low of 13% for aggravated assaults to a high of 55% for homicide (Langan & Solari, 1993). Figure 1.1 compares the odds of conviction for a violent crime in criminal versus juvenile courts (Jones & Krisberg, 1994). Other data also suggest that the sentencing of juveniles is not more lenient in juvenile courts compared to criminal courts. Data from California reveal that minors convicted and sentenced for violent crimes actually serve longer periods of incarceration in the California Department of Juvenile Justice than do adults who are sent to the state prison system (Jones & Krisberg, 1994).

Another study compares the sentences of 16- and 17-year-olds in New York and New Jersey. These two states have very different responses to youthful offenders. Whereas New York prosecutes most 16- and 17-year-olds in its criminal courts, New Jersey handles the vast majority of these youths in its juvenile courts. The researchers found that for youngsters who were accused of burglary and robbery, there were little differences in the severity of case dispositions in the juvenile courts of New Jersey compared to the criminal court proceedings in New York. Moreover, the study found that similar youths had lower rearrest rates if they were handled in the juvenile system rather than the adult court system (Fagan, 1991).

There is an impressive body of research that refutes the myth that nothing works with juvenile offenders. There are many studies showing the effectiveness of treatment responses for young offenders (Palmer, 1992). Gendreau and Ross (1987) have assembled an impressive array of studies showing the positive results of correctional interventions for juveniles. Others, such as

Figure 1.1   Odds of an Arrested Adult Being Convicted in Criminal Court vs. Odds of Conviction for Delinquency Referrals in 10 States

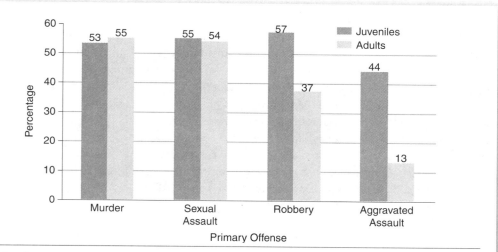

Juvenile Felony Defendants in Criminal Courts: Survey of 40 Counties, 1998
http://www.bjs.gov/content/pub/ascii/jfdcc98.txt
* In criminal court juveniles (64%) were more likely than adults (24%) to be charged with a violent felony.
SOURCE: Jones and Krisberg (1994, p. 25).

Greenwood and Zimring (1985) and Altschuler and Armstrong (1984), isolated the critical components of successful programs. More recently, Lipsey and Wilson (1998) and the National Council on Crime and **Delinquency** (NCCD, 2000) have summarized the promising treatment responses for serious and violent juvenile offenders. More details about these successful interventions are reviewed in Chapter 8. For now, it is important to see how the myth that juvenile offenders cannot be rehabilitated misguides policy changes that restrict the jurisdiction of the juvenile court and that impose harsher penalties on children.

A related myth that dominates political discourse on youth crime is that longer mandatory penalties of incarceration would reduce juvenile crime. This recommendation rests on the notion that there is a small number of offenders who are responsible for the vast majority of violent crime. Thus, if we could lock away these "bad apples" for a long period of time, the immediate crime problem would be greatly reduced. This idea has some natural intuitive appeal, since incarcerated offenders cannot commit offenses in the community. However, incarceration does not guarantee cessation of delinquent behavior. Further, there are several studies that point to the "dangerous few"—the small number of chronic offenders, in particular, gang members, who contribute to a disproportionate amount of violent crime (Loeber & Farrington, 2001).

There are several flaws in the argument that longer mandatory sentences would reduce violent youth crime. Presently, the vast majority of youngsters who are incarcerated in juvenile and adult correctional facilities have been convicted of nonviolent offenses (Jones & Krisberg, 1994). Broad-based policies mandating longer periods of confinement are most likely to increase the extent of incarceration for property offenders, drug offenders, and youths who are chronic minor offenders. Further, high-risk juvenile offenders do not remain high risk forever. Using **incapacitation** as a crime-control strategy assumes that criminal careers, once begun, will increase and include more violent behavior over time. Research studies on juvenile crime careers reveal a different picture. The prevalence of serious violent crime peaks between the ages of 16 and 17, and after age 20 the prevalence drops off sharply (Elliott, 1994). The likelihood that individuals will commit violent crimes during the ages of 21 to 27 is approximately the same as for children ages 12 and 13 (Elliott, 1994). Haapanen's (1988) long-term follow-up studies of youths released from the California Department of the Youth Authority show that longer sentences for youthful offenders would have little or no impact on overall societal rates of violent crime. In addition, the youth population is projected to increase substantially over the next 20 years. Thus, for every current juvenile offender that is taken out of circulation, there are increasing numbers entering their peak crime-committing years.

Juvenile justice professionals generally reject the notion that the incarceration system by itself can exert a major effect on reducing crime rates (NCCD, 1997). There is a growing body of knowledge that shows prevention and early intervention programs to be far more cost-effective than incapacitation in reducing rates of youth crime (Greenwood, Model, Rydell, & Chiesa, 1996). And there is some evidence that secure confinement of youngsters at early ages actually increases their subsequent offending behavior (Krisberg, 2016).

These are just some of the major myths that continue to confuse and confound the process of rational policy development for the juvenile justice system. There are others that claim to offer "miracle cures" for juvenile delinquency. Citizens are fed a regular diet of these miracle cures by the entertainment media and local news broadcasts, politicians, and entrepreneurs. Many communities have implemented programs such as Scared Straight that claimed that brief 1-day visits by youngsters to prison to be yelled at by inmates can cure emotional and family problems—despite compelling research that the program was not effective (Finckenaur, Gavin, Hovland, & Storvoll, 1999). Juvenile justice officials quickly jumped on the bandwagon to start up **boot camps** with scant evidence that these efforts could reduce **recidivism** (MacKenzie, 2000). Programs such as Tough Love and Drug Abuse Resistance Education have garnered media attention and significant funding without possessing solid empirical foundations. Perhaps the most destructive miracle cure to surface in recent years has been the mistaken belief that placing youths in adult prisons would advance public safety (Howell, 1998; Krisberg, 1997).

To achieve the ideal goal of juvenile justice, which is to protect vulnerable children and at the same time help build safer communities, these myths must be debunked. If the emperor indeed has no clothes, we need to acknowledge this fact and move to sounder social policies. The following chapters show how to assemble the evidence on which effective responses to youth crime can be built. We review the best research-based knowledge on what works and what does not. There is a path out of the morass of failed juvenile justice policies, but we must look critically at current policy claims, and we must apply high standards of scientific evidence to seek new answers.

Chapter 3 gives an important historical context to the ongoing quest for the juvenile justice ideal. The history of juvenile justice has not been a straightforward march to the more enlightened care of troubled youths. There have been race, class, and gender biases that have marked some of the contours of this history. Learning how we have arrived at the current system of laws, policies, and practices is a crucial step in conceiving alternatives to the status quo.

Figure 1.1 compares the odds of conviction for a violent crime in criminal vs. juvenile courts (Jones & Krisberg, 1994).

SUMMARY

Sadly, too much of what passes for public policy on juvenile justice has been founded on misinformation and mythology. Throughout our history, fear of the young, concern about immigrants, gender bias, and racial and class antagonism have dominated the evolution of juvenile justice. The media and politicians exploit these prejudices and fears for their own purposes.

The most powerful myth in the mid-1990s was the alleged wave of young superpredators. Some suggested that America would face an unprecedented increase in juvenile violence at their hands. This myth fueled a moral panic that shaped many public policies designed to get tougher with juvenile offenders. There were a large number of new laws that made it easier to try children in criminal courts and that increased the number of young people in prisons and jails. Critics of the juvenile court argued that it was too lenient in its sentencing practices, although there was little evidence backing these claims. Calls for longer periods of incarceration were also part of this moral panic. The research did not support the assertion that more incarceration would lead to lower juvenile crime rates.

REVIEW QUESTIONS

1. What was the superpredator myth? How did its proponents support their claims? Did the predicted juvenile crime wave occur, and if not, what did affect juvenile crime trends in the late 1990s?

2. What unsubstantiated claims are used to support harsher sentencing for juveniles?

3. Does increased incarceration reduce rates of juvenile crime? How has this conclusion been reached?

INTERNET RESOURCES

The Office of Juvenile Justice and Delinquency Prevention is the leading government resource on information and has the latest research on juvenile crime and the juvenile Justice system.
http://www.ojjdp.gov

The National Council on Crime and Delinquency is the oldest and most respected nonprofit research and policy group in the criminal justice and juvenile justice systems. The council is an excellent resource of research and policy options.
http://www.nccdglobal.org

The Center on Juvenile and Criminal Justice is the preeminent group on the current issues facing the juvenile justice system in California and many other states.
http://www.cjcj.org

Chapter 2

DATA SOURCES

Getting the facts straight about juvenile crime and about the operation of the juvenile justice system is a challenging job. One confronts a range of statistical data sources that are not integrated and that offer only very partial information on the subject. Further, America's justice systems are radically decentralized. There are more than 10,000 separate law enforcement agencies and nearly 3,000 juvenile courts. These agencies administer laws that are determined by each of the 50 states. In addition, juvenile justice systems in the District of Columbia, Native American tribal areas, and Puerto Rico, Guam, American Samoa, and American trust territories are governed by other legal structures administered by the federal government.

At the federal level, multiple agencies are responsible for collecting and analyzing statistical data on juvenile crime and juvenile justice. In various states and localities, accurate data are even more decentralized and difficult to obtain. The information about juvenile crime and the processing of young offenders that is collected on an ongoing basis is rarely, if ever, utilized for local planning or evaluation purposes. In this age of advanced information technology, the juvenile justice system continues to function and evolve based on anecdotes, subjective impressions, and the latest high-profile crime involving a young person that was reported on the evening news. This lack of reliable and timely data undoubtedly contributes to the ascendancy of myths in the formulation of juvenile justice policy and practice.

This chapter reviews available national sources of data and describes their strengths and weaknesses. Later, the chapter discusses a range of current research methods to deepen our understanding of the contours of youth crime and the juvenile justice system. Serious students of juvenile justice are best advised to master a broad range of research techniques to help them gain a more holistic comprehension of the subject. Students are also advised to be skeptical of all statements of fact and to become critical consumers of juvenile justice statistics and research findings.

UNDERSTANDING THE CONTOURS OF JUVENILE CRIME

It is a somewhat shocking statement, but no one knows for sure how much crime is attributable to young people or how many youngsters are crime victims. The U.S. Department of

Justice conducts an annual telephone survey to estimate the nature and extent of serious victimization, but these data do not tell us how much of this victimization is caused by young people. Moreover, the National Crime Victimization Survey (NCVS) does not collect data from children below the age of 12 because they believe that the results would be unreliable. This omission no doubt understates the extent of victimization that occurs in elementary schools, playgrounds, and in the home (Sickmund & Puzzanchera, 2014). Indeed, even for older adolescents, there are other national surveys suggesting that the NCVS underestimates the amount of criminal victimization experienced by teenagers (Wordes & Nuñez, 2002). Surveys of youths that are conducted in person, in homes, or at schools have produced estimates of teen victimization that are almost twice as high as the NCVS (Wordes & Nuñez, 2002).

Historically, citizens report only a fraction of their crime experiences to the police. Teens are even less likely to report being victimized, especially for violent offenses. Further, police make arrests in a small percentage of these cases ranging from 50% of reported violent crimes to 20% of property crimes. In the case of drug offenses, there is rarely a victim who voluntarily comes forward, so there are no valid estimates of how many drug offenses result in arrests.

The drop-off from reported victimization to official reports to the police and then to someone being arrested for a specific offense is highly significant for the understanding of juvenile crime. We know the age (or sex or ethnicity) of the offender only when he or she is taken into police custody. Thus, our main statistics about youth crime are, at best, descriptive of the subset of offenders who get caught. The relationship of this subset to the general population of juvenile offenders is unknown—and perhaps unknowable. Because the arrested group can be a very small proportion of those who are actually committing crimes, the basic statistical data and the nature and distribution of juvenile crime, or trends over time, are hopelessly inadequate. Further, those who are arrested are not necessarily guilty of the crime. In many jurisdictions, more than half of police referrals to the juvenile court result in dropped charges (Sickmund & Puzzanchera, 2014).

Limitations of Police Statistics on Juvenile Crime and Juvenile Justice

The odds of being arrested is more a measure of police practices, such as where police decide to patrol or to which crimes they pay the closest attention, rather than a valid indicator of youth behavior. This discrepancy is especially true for crimes such as drug trafficking, in which both the offender and other crime participants may want to keep the matter low profile. Huizinga and Elliott (1987) have compared the numbers of crimes that were reported by juveniles in confidential surveys with the number of these offenses that resulted in an official arrest. They found that only about 1 in 7 very serious crimes that were admitted to in the confidential surveys resulted in an arrest; for less serious crimes, the odds of getting caught were 1 in 50. This discrepancy casts serious doubt on research and evaluation generalizations that are made using only police statistics. Later in this chapter we discuss the utility and limitations of self-report surveys to advance our understanding of the contours of youth crime.

National data on juvenile arrests are assembled by the Federal Bureau of Investigation from reports submitted by thousands of local law enforcement agencies through the Uniform Crime Reporting System (UCR). From year to year, the number of participating agencies varies, making analyses of trends extremely problematic. For example, for several years in the 1990s, the Chicago Police Department did not report data on juvenile arrests to the Illinois Criminal Justice Authority, which compiles statewide crime statistics. The omission of Chicago data created an obvious distortion in statistics pertaining to Cook County (where Chicago is one of a number of cities) and for overall Illinois juvenile crime data (Krisberg, Noya, Jones, & Wallen, 2001). Law enforcement reporting is a little more consistent for the most serious offenses—including homicide, rape, robbery, burglary, larceny, auto theft, and arson—but the data from various jurisdictions are not at all uniform for minor assaults—vandalism, drug offenses, and **status offenses** such as truancy, running away, curfew violations, possession of alcohol, and incorrigibility. These latter offenses are among the most frequent crimes resulting in teenage arrests. Thus, juvenile crime data are extremely uneven and frustrate the ability to plan and evaluate responses to juvenile crime at the federal, state, and local levels.

Even less is known, in any systematic manner, about how the police handle youths once they are arrested. The annual UCR report, *Crime in the United States,* contains just one table that provides data on police dispositions of juvenile arrests. Although the information is broken down by the size of the jurisdiction, the data are not presented for specific cities or states. In general, it appears that law enforcement agencies refer about two thirds of arrests to the juvenile court. A very small fraction of juvenile arrests (roughly 7%) are directly referred to criminal courts. The remaining juvenile arrests are handled informally within the police department, referred to community welfare agencies, and about 67% of arrests are referred to family or juvenile courts. Local police practices are extremely diverse in terms of the methods used to informally process juvenile arrestees, ranging from giving the youth a referral card to more formal counseling or informal supervision programs operated within law enforcement agencies (sometimes in partnership with other community agencies). Some police agencies favor informal handling of juvenile cases, whereas others emphasize much more formal case processing through custody. This variability in police practices limits the ability to utilize police data to compare policies or programs between jurisdictions or within the same locale over a long time period. The local media is fond of publicizing year-to-year trends in juvenile arrests, but these comparisons are rarely accurate reflections of actual youthful offending.

JUVENILE COURT STATISTICS

Data on the operation of the juvenile court were first collected by the U.S. Children's Bureau in 1926. These data were in summary form, permitting very little in the way of detailed analyses. With support from the federal Office of Juvenile Justice and Delinquency Prevention (OJJDP), the quality and quantity of juvenile court statistics was greatly enhanced. Compiled by the National Center for Juvenile Justice (NCJJ), aggregate data are collected by more than 80% of the nation's juvenile court jurisdictions (Stahl, 1998). These data cover the total number of cases handled, a distribution of cases by gender, and whether each case was handled without a formal delinquency **petition**. Beginning in 1979, NCJJ began collecting individual-level data from 15 states or large jurisdictions with automated data systems. Today, these individual-level data come from juvenile courts that represent 83% of the youth population and are referred to as the National Juvenile Court Data Archives (Sickmund & Puzzanchera, 2014).

As with police arrest statistics, juvenile court data analysis confronts a broad range of state laws and local practices that frustrate cross-jurisdictional comparisons. Further, since the NCJJ data are not based on a systematic sample of all jurisdictions (they include only the reporting courts), any national estimates or trends over time must be viewed with extreme caution. Still, some data are better than none, and the National Juvenile Court Data Archives are a rich source of research material on the operation of the juvenile court.

There are no existing national or state data sources covering the practices of prosecutors in the juvenile court. The Juvenile Defendants in Criminal Court Program (JDCC) of the Pretrial

Services Resource Center does provide some prosecution data on juveniles tried in adult courts in approximately 40 of the nation's 75 most populous jurisdictions (Henry, Moffitt, Robins, Earls, & Silva, 1993).

JUVENILES TAKEN INTO CUSTODY

Data on juvenile corrections are more problematic than the court data. Beginning in the early 1970s, the U.S. Bureau of the Census administered a biannual survey to all public and private juvenile correctional facilities. Known popularly as the Children in Custody (CIC) Survey, this was the primary source of information on children in confinement for a quarter of a century (Smith, 1998). There were always concerns that the survey did not adequately cover all types of juvenile facilities, and the growth of correctional programs operated by private agencies made this problem more difficult. The Bureau of the Census was able to get a near 100% return on surveys sent to public juvenile facilities. Like most national surveys, CIC was hampered by the same problems of differing legal codes that governed youth corrections in the states (Dedel, 1998). Data were reported to the Bureau of the Census based on aggregate counts, as opposed to information on individual residents, limiting the extent of data and policy analysis that could be accomplished. Though CIC information based on a 1-day snapshot of the incarcerated population seemed reasonably reliable, other data that included estimates of annual admissions or length of stay were far less credible.

There have been a variety of attempts to supplement the CIC. In 2000, the Bureau of Justice Statistics launched an in-depth survey of a **random sample** of youths in secure facilities (Beck, Kline, & Greenfield, 1988) that is currently being replicated by Westat (U.S. Bureau of Justice Statistics, 2013). This National Survey of Youth in Custody provides a more intensive view of the kinds of youth in secure facilities and the personal issues that they have confronted in their lives. The current Westat replication will cover far more topics, including the **conditions of confinement**.

In the 1990s, the National Council on Crime and Delinquency (NCCD) field-tested and then implemented a data collection effort designed to gather basic information on youths who were admitted annually to state-operated commitment facilities. These data were individual-based, permitting richer policy analysis and allowing for more precise calculations of length of stay (DeComo, Tunis, Krisberg, & Herrera, 1993). More than 60,000 admissions were captured in 1996. The primary limitation of this NCCD effort was that not all states participated (similar to the National Juvenile Court Data Archives), and the data collection did not cover short-term **detention** facilities that are mostly run by units of local government, into which the vast majority of youths were taken into custody. The NCCD state-level data collection did capture the vast majority of very serious offenders in juvenile corrections, as well as those who were confined for the longest periods of time. This problem is quite similar to those encountered by researchers in the field of adult corrections. There are reasonably complete data on persons entering state prison systems, but we know far less about the attributes of the much larger pool of individuals sent to local jails.

After 1995, OJJDP retired both the traditional CIC survey and the NCCD effort to get admissions and releases from state juvenile facilities. Instead, OJJDP funded the Bureau of the Census to complete a 1-day roster of all youths in facilities covered by the previous CIC survey. The roster approach was intended to collect individual-level data on youths in a wide variety of custodial settings (Moone, 1997). This approach failed to collect data on length of stay or estimates of admissions to juvenile facilities. The first roster was completed in 1997 and the second in 2000.

The growing practice of trying children in criminal courts requires that researchers and policymakers examine data sets on adult corrections to develop a complete portrait of juvenile incarceration. Very limited data are available on this key aspect of the justice system. For example, the National **Jail** Survey, conducted by the U.S. Bureau of Justice Statistics (BJS), provides periodic counts of the number of juveniles admitted to jails, as well as a 1-day count of residents in these facilities. Another BJS data collection effort, the National Correctional Reporting Program, has data from several states on the numbers of persons under age 18 who enter prisons in the reporting states. In both these statistical data sources, the definition of a juvenile is derived from the local jurisdiction, thus perpetuating the problem of wide variability in the sorts of data that are reported from differing locales.

SELF-REPORTED DELINQUENCY SURVEYS

The obvious limitations in official justice system data sources have led a number of researchers to base their analyses of youth crime on surveys that are either conducted in school settings or in the homes of youths. These surveys ask young people to report about the incidence and prevalence of their misconduct. Such self-report surveys have been utilized since the mid-1970s. Increasingly, researchers examining the causes and correlates of delinquent behavior, drug use, and sexual behavior rely on self-report surveys as their primary source of information. Two of the most influential of these surveys are the National Youth Survey (Elliott & Ageton, 1980) and the Monitoring the Future Project (Johnson & Bechman, 1981). Sponsored by the National Institute of Mental Health, the National Youth Survey, begun in 1976, was designed to gain a better understanding of both conventional and deviant types of behavior by youths. It involved collecting information from a representative sample of young people in the United States. Seven waves of the survey were conducted—1976, 1977, 1978, 1979, 1980, 1983, and 1987. The Monitoring the Future Project is an ongoing annual study of the behaviors, attitudes, and values of American secondary school students, college students, and young adults. Each year, approximately 50,000 eighth-, 10th-, and 12th-grade students are surveyed (12th graders since 1975, and eighth and 10th graders since 1991). In addition, annual follow-up questionnaires are mailed to a sample of each graduating class for a number of years after their initial participation.

In general, self-report surveys reveal that a much larger proportion of the youth population admits to engaging in law-breaking behavior—a far greater number of offenders than are apprehended by the police. Further, self-report surveys cast significant doubts on the facts about delinquency derived from studies of captured offenders. For example, the large differences among ethnic groups, age-specific rates of offending, or different crime patterns among boys and girls that are observed in police or court data often are diminished or change radically in the results of the self-report studies (Elliott, 1994).

Critics of self-report surveys have raised questions about the willingness of various groups of young people to honestly report on their criminal behavior. Some youngsters may fail to disclose their misconduct, whereas others will brag or exaggerate it. It is reasonable to assume that some behaviors, including sex offending, may be subject to underreporting. Concerns have also been expressed that self-report studies contain a large amount of very minor law violations, whereas official statistics may reflect more serious offenses. Another concern with self-report surveys is the dependency on English language skills that may lead to distorted data from youngsters for whom English is a second language.

QUALITATIVE DATA SOURCES

So far, the discussion about data sources has mainly focused on quantitative and statistical information. These sorts of numerical data are very useful in understanding the contours of delinquent behavior and the operations of the juvenile justice system. But quantitative data can provide only a narrow comprehension of the subject. Qualitative methods are needed to achieve a holistic understanding of the subject matter—what the great sociologist Max Weber called *verstehen*.

There are many potential sources of humanistic data on juvenile delinquency and juvenile justice. For example, the work of scholars at the University of Chicago during the 1920s provided some of the richest sources of data on urban life. Most notably, Frederick Thrasher's *The Gang* (1927) provided a crucial understanding of adolescent behavior that remains fresh and relevant today. Another famous Chicago sociologist, Clifford Shaw, employed the technique of a delinquent's own story, a kind of sociological biography, in such classics as *Brothers in Crime* (1938) and *The Jack-Roller: A Delinquent Boy's Own Story* (1930). Shaw provided vital descriptions of underworld life that fueled the growth of criminological theory in the 1930s and 1940s.

Social scientists employing the delinquent's own perspective to comprehend the world of youth crime continue to the present period. Peterson and Truzzi (1972) offer a splendid collection of edited interviews with offenders. Krisberg (1971) uses observations and interviews with Philadelphia gang members to give us a glimpse of their worldviews. Other gang researchers, such as Moore and Hagedorn (1988) and Sanchez-Jankowski (1991), offer very useful insights into juvenile delinquency through extensive interviews and their descriptions of hanging out with gang members. Jack Katz, in *Seductions of Crime: Moral and Sensual Attractions in Doing Evil* (1988), uses qualitative methods to supplement quantitative data. Katz argues that traditional social science methods provide excellent views of the background of crime but that more humanistic data are needed to adequately describe the foreground, the proximate forces that propel individuals into criminal behavior.

In addition to the works of sociologists of the street corner, the student of juvenile delinquency can utilize a broad range of autobiographical materials that portray the lives of youthful offenders. These can take the form of both fiction and nonfiction, such as Alex Haley's *The Autobiography of Malcolm X* (1965), Claude Brown's *Manchild in the Promised Land* (1966), Piri Thomas's *Down These Mean Streets* (1967), or the more recent autobiography of a Los Angeles gang member in Sanyika Shakur's *Monster* (1993). Likewise, there are a number of theatrical productions and films that can offer a vivid picture of the lives of youthful offenders, such as *Rebel without a Cause* (1955), *The Education of Sonny Carson* (1974), *Boyz N the Hood* (1991), *American Me* (1992), or the classic *Angels With Dirty Faces* (1938). The student can also look to some lyrics from contemporary urban music such as hip-hop and rap for further understanding of the worlds of delinquent youths.

Qualitative methods are also very useful for theory building about the juvenile justice system. Observations studies such as those conducted by Wilson (1968), Cicourel (1968), and Emerson (1969) provide very valuable analyses of the norms and operations of the juvenile justice system in a variety of communities. Autobiographical works beginning with Charles Loring Brace's *The Dangerous Classes of New York* (1872), Jane Addams's *Twenty Years at Hull-House* (1911), and Judge Ben Lindsey's *The Dangerous Life* (1931) offer very fascinating accounts of the evolution of the juvenile justice system in America. Later works by juvenile court judges such as Lois Forer (1970) and Lisa Richette (1969) offer a more contemporary view of the juvenile court. There are also a number of excellent accounts of juvenile justice systems by journalists (Hubner & Wolfson, 1996) that can deepen the student's learning about the workings of the justice system, and fascinating personal accounts of reform campaigns such as Jerome Miller's important story of reforming juvenile corrections in Massachusetts in *Last One Over the Wall* (1998).

SUMMARY

There are several sources of data on juvenile offending. None of the available data is comprehensive, and they all have significant limitations. Information on juvenile crime can be gleaned from arrest statistics, juvenile court data, information on youth in custody, self-report surveys, and a range of qualitative sources. Much of the existing research is based on arrest statistics, but the proportion of those arrested is a small, and probably biased, sample of all those youngsters committing crimes. Moreover, the highly decentralized nature of the juvenile justice system, including significant differences in law and practice among communities, makes interpreting available official data quite challenging. Any conclusions drawn from existing statistical systems must be treated with some caution. Given limitations in each data source, researchers and policymakers are encouraged to utilize multiple sources to complement their view of juvenile crime.

REVIEW QUESTIONS

1. What are some of the major obstacles to gaining an accurate statistical picture of juvenile crime and juvenile justice?

2. Compare and contrast the relative strengths and weaknesses of self-report data, official arrest statistics, and qualitative data to guide our understanding of juvenile crime and juvenile justice.

INTERNET RESOURCES

The U.S. Bureau of Justice Statistics is the largest federal compilation of data on crime, victimization, and the justice system.
http://www.bjs.gov

The National Center for Juvenile Justice is the research arm of the National Council of Family and Juvenile Court Judges. This group produces a regular report that summarizes almost all of the available national and state data on juvenile offending, victimization, and the juvenile justice system.
www.ncjj.org

Chapter 3

THE HISTORICAL LEGACY OF JUVENILE JUSTICE

The first institution for the control of juvenile delinquency in the United States was the New York **House of Refuge**, founded in 1825, but specialized treatment of wayward youth has a much longer history—one tied to changes in the social structure of medieval Europe. These same changes prompted the colonization of the New World and led to attempts to control and exploit the labor of African, European, and Native American children.

Virtually all aspects of life were in a state of flux for the people of Europe in the later Middle Ages (16th and 17th centuries). The economy was being transformed from a feudal system based on sustenance agriculture to a capitalistic, trade-oriented system focusing on cash crops and the consolidation of large tracts of land. In religious matters, the turmoil could be amply witnessed in the intense struggles of the Reformation. Politically, power was increasingly concentrated in the hands of a few monarchs, who were fashioning strong centralized states. The growth of trade and exploration exposed Europeans to a variety of world cultures and peoples.

For the lower classes of European society, these were "the worst of times." The rising population density as well as primitive agricultural methods led to a virtual exhaustion of the land. Increasing urban populations created new demands for cheap grain, and landlords responded by increasing the fees paid by peasants who worked the land. Large numbers of peasants were displaced from the land to permit the growth of a capitalist pasturage system. The standard of living of the European peasantry dropped sharply, and this new, displaced class streamed into the cities and towns in search of means of survival. The workers and artisans of the cities were deeply threatened by the prospect that this pauper class would drive down the general wage level. Most European towns experienced sharp rises in crime, rioting, and public disorder.

To control and defuse the threat of this new "dangerous class," the leaders of the towns enacted laws and other restrictions to discourage immigration and contain the movement of the impoverished peasantry. "Poor laws" were passed, preventing the new migrants from obtaining citizenship, restricting their membership in guilds, and often closing the city gates to them. Vagrancy laws were instituted to control and punish those who seemed a threat to the social order. Certain legislation, such as the Elizabethan Statute of Artificers (1562), restricted access into certain trades, forcing the rural young to remain in the countryside.

Urban migration continued despite most attempts to curtail it. The collective units of urban life, the guild and the family, began to weaken under the pressure of social change. Children often were abandoned or released from traditional community restraints. Countless observers from the period tell of bands of youths roaming the cities at night, engaging in thievery, begging, and other forms of misbehavior (Sanders, 1970).

At this time, family control of children was the dominant model for disciplining wayward youth. The model of family government, with the father in the role of sovereign, was extended to those without families through a system of *binding out* the young to other families. Poor children, or those beyond parental control, were apprenticed to householders for a specified period of time. Unlike the apprenticeship system for the privileged classes, the binding-out system did not oblige the master to teach his ward a trade. Boys generally were assigned to farming tasks and girls were brought into domestic service.

As the problem of urban poverty increased, the traditional modes of dealing with delinquent or destitute children became strained. Some localities constructed institutions to control wayward youth. The Bridewell (1555) in London is generally considered the first institution specifically designed to control youthful beggars and vagrants. In 1576, the English Parliament passed a law establishing a similar institution in every English county. The most celebrated of these early institutions was the Amsterdam House of Corrections (1595), which was viewed as an innovative solution to the crime problem of the day.[1] The houses of correction combined the principles of the poorhouse, the workhouse, and the penal institution. The youthful inmates were forced to work within the institution and thus develop habits of industriousness. Upon release, they were expected to enter the labor force, so house of correction inmates often were hired out to private contractors. Males rasped hardwoods used in the dyeing industry, and when textile manufacturing was introduced to the houses of correction, this became the special task of young woman inmates.

The early houses of correction, or so-called "Bridewells," accepted all types of children including the destitute, the infirm, and the needy. In some cases, parents placed their children in these institutions because they believed the regimen of work would have a reformative effect. Although it is debatable whether the houses of correction were economically efficient, the founders of such institutions clearly hoped to provide a cheap source of labor to local industries. The French institutions, called *hospitaux generaux*, experimented with technological improvements and different labor arrangements. This often brought charges of unfair competition from guilds, who feared the demise of their monopoly on labor, and businessmen, who felt threatened by price competition at the marketplace. Some authors stress the economic motive of these early penal institutions: "The institution of the houses of correction in such a society was not the result of brotherly love or of an official sense of obligation to the distressed. It was part of the development of capitalism" (Rusche & Kirchheimer, 1939, p. 50).

The enormous social, political, and economic dislocations taking place in Europe provided a major push toward colonization of the Americas. People emigrated for many reasons—some to get rich, some to escape political or religious oppression, and some because they simply had nothing to lose. Settlement patterns and the resulting forms of community life varied considerably. In the Massachusetts Bay Colony, for example, the Puritans attempted to establish a deeply religious community to serve God's will in the New World. The Puritans brought families with them and from the outset made provisions for the care and control of youths.

In contrast, the settlement of Virginia was more directly tied to economic considerations. There were persistent labor shortages, and the need for labor prompted orders for young people to be sent over from Europe. Some youths were sent over by "spirits," who were agents of merchants or ship owners. The spirits attempted to persuade young people to immigrate to America. They often promised that the New World would bring tremendous wealth and happiness to the youthful immigrants. The children typically agreed to work a specific term (usually 4 years) in compensation for passage across the Atlantic and for services rendered during the trip. These agreements of service were then sold to inhabitants of the new colonies, particularly in the South. One can imagine that this labor source must have been quite profitable for the plantations of the New World. Spirits were often accused of kidnapping, contractual fraud, and deception of a generally illiterate, destitute, and young clientele.

Other children coming to the New World were even more clearly coerced. For example, it became an integral part of penal practice in the early part of the 18th century to transport prisoners to colonial areas. Children held in the overcrowded Bridewells and poorhouses of England were brought to the Americas as indentured servants. After working a specified number of years as servants or laborers, the children were able to win their freedom. In 1619, the colony of Virginia regularized an agreement for the shipment of orphans and destitute children from England.

That same year, Africans, another group of coerced immigrants, made their first appearance in the Virginia Colony. The importation of African slaves eventually displaced the labor of youthful poor because of greater economic feasibility. The black chattels were physically able to perform strenuous labor under extreme weather conditions without adequate nutrition. These abilities would finally be used to describe them as beasts. Also, the high death rates experienced under these conditions did not have to be accounted for. The bondage of Africans was soon converted into lifetime enslavement, which passed on through generations. The southern plantation system, dependent on the labor of African slaves, produced tremendous wealth, further entrenching this inhuman system (Stamp, 1956; Yetman, 1970). Racism, deeply lodged in the English psyche, provided the rationale and excuse for daily atrocities and cruelties.[2]

Studies of slavery often overlook the fact that most slaves were children (Ward, 2012). Slave traders thought children would bring higher prices. Accounts of the slave trade emphasize the economic utility of small children, who could be jammed into the limited cargo space available on slave ships. Children were always a high proportion of the total slave population because slave owners encouraged the birth of children to increase their capital. Little regard was paid by slave owners to keeping families together. African babies were a commodity to be exploited just as one might exploit the land or the natural resources of a plantation, and young slave women often were used strictly for breeding. A complete understanding of the social control of children must include a comparison of the institution of slavery to the conditions faced by children in other sections of the country.[3]

Another group of children who often are ignored in discussions of the history of treatment of youth in North America are Native Americans. In 1609, officials of the Virginia Company were authorized to kidnap Native American children and raise them as Christians. The stolen youths were to be trained in the religion, language, and customs of the colonists. The early European colonists spread the word of the Gospel to help rationalize their conquests of lands and peoples. But an equally important motivation was their interest in recruiting a group of friendly natives to assist in trade negotiations and pacification programs among the native peoples. The early Indian schools resembled penal institutions, with heavy emphasis on useful work, Bible study, and religious worship. Although a substantial amount of effort and money was invested in Indian schools, the results were considerably less than had been originally hoped:

> Missionaries could rarely bridge the chasm of mistrust and hostility that resulted from wars, massacres and broken promises. With so many colonists regarding the Indian as the chief threat to their security and the Indians looking upon the colonists as hypocrites, it is little wonder that attempts to win converts and to educate should fail. (Bremner, Barnard, Hareven, & Mennel, 1970, p. 72)

Unlike attempts to enslave children of African descent, early efforts with Native Americans were not successful. Relations between European colonists and Native Americans during this period centered on trading and the securing of land rights (Chavez-Garcia, 2012). These contrasting economic relationships resulted in divergent practices in areas such as education. Although there was general support for bringing "the blessings of Christian education" to the Native American children, there was intense disagreement about the merits of educating African slaves. Whereas some groups, such as the Society for the Propagation of the Gospel, argued that all "heathens" should be educated and converted, others feared that slaves who were baptized would claim the status of free men. There was concern among whites that education of slaves would lead to insurrection and revolt. As a result, South Carolina and several other colonies proclaimed that conversion to Christianity would not affect the status of slaves (Bremner et al., 1970). Many southern colonies made it a crime to teach reading and writing to slaves. A middle-ground position evolved, calling for religious indoctrination without the more dangerous education in literacy (Bremner et al., 1970; Gossett, 1963).

In the early years of colonization, the family was the fundamental mode of juvenile social control, as well as the central unit of economic production. Even in situations where children were apprenticed or indentured, the family still served as the model for discipline and order (Mintz, 2004). Several of the early colonies passed laws requiring single persons to live with families. The dominant form of poor relief at this time was placing the needy with other families in the community (Rothman, 1971). A tradition of family government evolved in which the father was empowered with absolute authority over all affairs of the family. Wives and children were expected to give complete and utter obedience to the father's wishes. This model complemented practices in political life, where absolute authority was thought to be crucial to the preservation of civilization.

Colonial laws supported and defended the primacy of family government. The earliest laws concerning youthful misbehavior prescribed the death penalty for children who disobeyed their parents. For example, part of the 1641 Massachusetts *Body of Liberties* reads as follows:

> If any child, or children, above sixteen years of age, and of sufficient understanding, shall CURSE or SMITE their natural FATHER or MOTHER, he or they shall be putt to death, unless it can be sufficiently testified that the Parents have been very unchristianly negligent in the education of such children: so provoked them by extreme and cruel correction, that they have been forced thereunto, to preserve themselves from death or maiming: *Exod* 21:17, *Lev* 20:9, *Exod* 21:15. (Hawes, 1971, p. 13)

Although there is little evidence that children were actually put to death for disobeying their parents, this same legal principle was used to justify the punishment of rebellious slave children in the southern colonies. Family discipline typically was maintained by corporal punishment. Not only were parents held legally responsible for providing moral education for their children, but a Massachusetts law of 1642 also mandated that parents should teach their children reading and writing. Later, in 1670, public officials called *tithing men* were assigned to assist the selectmen (town councilmen) and constables in supervising family government. The tithing men visited families who allegedly were ignoring the education and socialization of their children. Although there are records of parents brought to trial due to their neglect of parental duties, this manner of supervising family government was not very successful.

The family was the central economic unit of colonial North America. Home-based industry, in which labor took place on the family farm or in a home workshop, continued until the end of the 18th century. Children were an important component of family production, and their labor was considered valuable and desirable. A major determinant of a child's future during this time was the father's choice of apprenticeship for his child. Ideally, the apprenticeship system was to be the stepping stone into a skilled craft, but this happy result was certain only for children of the privileged classes. As a consequence, children of poor families might actually be *bound out* as indentured servants. The term of apprenticeship was generally 7 years, and the child was expected to regard his master with the same obedience due natural parents. The master was responsible for the education and training of the young apprentice, and he acted *in loco parentis,* assuming complete responsibility for the child's material and spiritual welfare. Although apprenticeships were voluntary for the wealthier citizens, for the wayward or destitute child they were unavoidable. The use of compulsory apprenticeships was an important form of social control exercised by town and religious officials upon youths perceived as troublesome (Bremner et al., 1970).

The industrial revolution in North America, beginning at the end of the 18th century, brought about the gradual transformation of the labor system of youth. The family-based productive unit gave way to an early factory system. Child labor in industrial settings supplanted the apprenticeship system. As early as the 1760s, there were signs that the cotton industry in New England would transform the system of production, and by 1791, all stages in the manufacture of raw cotton into cloth were performed by factory machinery. The Samuel Slater factory in Providence, Rhode Island, employed 100 children aged 4 to 10 years in cotton manufacture. Here is a description of the workplace environment:

> They worked in one room where all the machinery was concentrated under the supervision of a foreman, spreading the cleaned cotton on the carding machine to be combed and passing it though the roving machine, which turned the cotton into loose rolls ready to be spun. Some of the children tended the spindles, removing and attaching bobbins. Small, quick fingers were admirably suited for picking up and knotting broken threads. To the delight of Tench Coxe, a champion of American industry, the children became "the little fingers . . . of the gigantic automatons of labor-saving machinery." (Bremner et al., 1970, p. 146)

During the next two decades, the use of children in New England industrial factories increased, and children comprised 47% to 55% of the labor force in the cotton mills. The proliferation of the factory system transformed the lives of many Americans. On one hand, enormous wealth began to accumulate in the hands of a few individuals. At the same time, the switch from a family-based economy to a factory system where workers sold their labor meant that many families were displaced from the land. A large class of permanently impoverished Americans evolved. The use of child labor permitted early industrialists to depress the general wage level. Moreover, companies provided temporary housing and supplies to workers at high prices, so that workers often incurred substantial debts rather than financial rewards.

Increased child labor also contributed to the weakening of family ties because work days were long and often competed with family chores. Children were now responsible to two masters—their fathers and their factory supervisors. Work instruction became distinct from general education and spiritual guidance as the family ceased to be an independent economic unit. Conditions of poverty continued to spread, and the social control system predicated upon strong family government began to deteriorate. During the first decades of the 19th century, one could begin to observe a flow of Americans from rural areas to the urban centers. As increasing economic misery combined with a decline in traditional forms of social control, an ominous stage was being set. Some Americans began to fear deeply the growth of a "dangerous class" and attempted to develop new measures to control the wayward youth who epitomized this threat to social stability.

THE HOUSES OF REFUGE (1825–1860)[4]

Severe economic downturns in the first two decades of the 19th century forced many Americans out of work. At the same time, increasing numbers of Irish immigrants arrived in the United States. These changes in the social structure, combined with the growth of the factory system, contributed to the founding of specialized institutions for the control and prevention of juvenile delinquency in the United States (Hawes, 1971; Mennel, 1973; Pickett, 1969).

As early as 1817, the more privileged Americans became concerned about the apparent connection between increased pauperism and the rise of delinquency. The Society for the Prevention of Pauperism was an early attempt to evaluate contemporary methods of dealing with the poor and to suggest policy changes. This group also led campaigns against taverns and theaters, which they felt contributed to the problem of poverty. The efforts of several members of this group led to the founding in New York City of the first House of Refuge in 1825. The group conducted investigations, drew up plans and legislation, and lobbied actively to gain acceptance of their ideas. In other northeastern cities, such as Boston and Philadelphia, similar efforts were under way.

A number of historians have described these early 19th-century philanthropists as "conservative reformers" (Coben & Ratner, 1970; Mennel, 1973). These men were primarily from wealthy, established families and often were prosperous merchants or professionals. Ideologically, they were close to the thinking of the colonial elite, and later, to the Federalists. Popular democracy was anathema to them because they viewed themselves as God's elect and felt bound to accomplish his charitable objectives in the secular world. Leaders of the movement to establish the houses of refuge, such as John Griscom, Thomas Eddy, and John Pintard, viewed themselves as responsible for the moral health of the community, and they intended to regulate community morality through the example of their own proper behavior as well as through benevolent activities. The poor and the deviant were the objects of their concern and their moral stewardship.

Although early 19th-century philanthropists relied on religion to justify their good works, their primary motivation was the protection of their class privileges. Fear of social unrest and chaos dominated their thinking (Mennel, 1973). The rapid growth of a visible impoverished class, coupled with apparent increases in crime, disease, and immorality, worried those in power. The bitter class struggles of the French Revolution and periodic riots in urban areas of the United States signaled danger to the status quo. The philanthropy of this group was aimed at reestablishing social order, while preserving the existing property and status relationships. They were responsible for founding such organizations as the American Sunday School Union, the American Bible Society, the African Free School Society, and the Society for Alleviating the Miseries of

Public Prisons. They often were appointed to positions on boards of managers for lunatic asylums, public hospitals, workhouses for the poor, and prisons.

The idea for houses of refuge was part of a series of reform concepts designed to reduce juvenile delinquency. Members of the Society for the Prevention of Pauperism were dissatisfied with the prevailing practice of placing children in adult jails and workhouses. Some reformers felt that exposing children to more seasoned offenders would increase the chances of such children becoming adult criminals. Another issue was the terrible condition of local jails. Others worried that, due to these abominable conditions, judges and juries would lean toward acquittal of youthful criminals to avoid sending them to such places. Reformers also objected that the punitive character of available penal institutions would not solve the basic problem of pauperism. The reformers envisioned an institution with educational facilities, set in the context of a prison. John Griscom called for "the erection of new prisons for juvenile offenders" (Mennel, 1973). A report of the Society for the Prevention of Pauperism suggested the following principles for such new prisons:

> These prisons should be rather schools for instruction, than places of punishment, like our present state prisons where the young and the old are confined indiscriminately. The youth confined there should be placed under a course of discipline, severe and unchanging, but alike calculated to subdue and conciliate. A system should be adopted that would provide a mental and moral regimen. (Mennel, 1973, p. 11)

By 1824, the society had adopted a state charter in New York under the name of the Society for the Reformation of Juvenile Delinquents and had begun a search for a location for the House of Refuge.

On New Year's Day 1825, the New York House of Refuge opened with solemn pomp and circumstance. A year later, the Boston House of Reformation was started, and in 1828, the Philadelphia House of Refuge began to admit wayward youth. These new institutions accepted both children convicted of crimes and destitute children. Because they were founded as preventive institutions, the early houses of refuge could accept children who "live an idle or dissolute life, whose parents are dead or if living, from drunkenness, or other vices, neglect to provide any suitable employment or exercise any salutary control over said children" (Bremner et al., 1970, p. 681). Thus, from the outset, the first special institutions for juveniles housed together delinquent, dependent, and **neglected children**—a practice still observed in most juvenile detention facilities today.[5]

The development of this new institution of social control necessitated changes in legal doctrines to justify the exercise of power by refuge officials. In *Commonwealth v. M'Keagy* (1831), the Pennsylvania courts had to rule on the legality of a proceeding whereby a child was committed to the Philadelphia House of Refuge on the weight of his father's evidence that the child was "an idle and disorderly person." The court affirmed the right of the state to take a child away from a parent in cases of vagrancy or crime, but because this child was not a vagrant, and the father was not poor, the court ruled that the child should not be committed. Judicial officials did not wish to confuse protection of children with punishment because this might engender constitutional questions as to whether children committed to houses of refuge had received the protection of due process of law.

The related question of whether parental rights were violated by involuntary refuge commitments was put to a legal test in *Ex parte Crouse* (1838). The father of a child committed to the Philadelphia House of Refuge attempted to obtain her release through a **writ** of **habeas corpus.**

The state supreme court denied the motion, holding that the right of parental control is a natural but not inalienable right:

> The object of the charity is reformation, by training the inmates to industry; by imbuing their minds with principles of morality and religion; by furnishing them with means to earn a living; and, above all, by separating them from the corrupting influence of improper associates. To this end, may not the natural parents, when unequal to the task of education, or unworthy of it, be superseded by the **parens patriae**, or common guardian of the community? The infant has been snatched from a course which must have ended in confirmed depravity; and, not only is the restraint of her person lawful, but it would have been an act of extreme cruelty to release her from it. (*Ex parte Crouse*, 1838)

The elaboration of the doctrine of **parens patriae** in the *Crouse* case was an important legal principle used to support the expanded legal powers of the juvenile court. It is important to recognize the significance of both social class and hostility toward Irish immigrants in the legal determination of the *Crouse* case.[6] Because Irish immigrants were viewed at this time as corrupt and unsuitable as parents, it is easy to see how anti-immigrant feelings could color judgments about the suitability of parental control. As a result, children of immigrants made up the majority of inmates of the houses of refuge.

The early houses of refuge either excluded blacks or housed them in segregated facilities. In 1849, the city of Philadelphia opened the House of Refuge for Colored Juvenile Delinquents. Racially segregated refuges were maintained in New York City and Boston only through the limited funds donated by antislavery societies. Because refuge managers viewed all young women delinquents as sexually promiscuous with little hope for eventual reform, young women also received discriminatory treatment.[7]

The managers of houses of refuge concentrated on perfecting institutional regimens that would result in reformation of juveniles. Descriptions of daily activities stress regimentation, absolute subordination to authority, and monotonous repetition:

> At sunrise, the children are warned, by the ringing of a bell, to rise from their beds. Each child makes his own bed, and steps forth, on a signal, into the Hall. They then proceed, in perfect order, to the Wash Room. Thence they are marched to parade in the yard, and undergo an examination as to their dress and cleanliness; after which they attend morning prayer. The morning school then commences, where they are occupied in summer, until 7 o'clock. A short intermission is allowed, when the bell rings for breakfast; after which, they proceed to their respective workshops, where they labor until 12 o'clock, when they are called from work, and one hour allowed them for washing and eating their dinner. At one, they again commence work, and continue at it until five in the afternoon, when the labors of the day terminate. Half an hour is allowed for washing and eating their supper, and at half-past five, they are conducted to the school room, where they continue at their studies until 8 o'clock. Evening Prayer is performed by the Superintendent; after which, the children are conducted to their dormitories, which they enter, and are locked up for the night, when perfect silence reigns throughout the establishment. The foregoing is the history of a single day, and will answer for every day in the year, except Sundays, with slight variations during stormy weather, and the short days in winter. (Bremner et al., 1970, p. 688)[8]

Routines were enforced by corporal punishment as well as other forms of control. Houses of refuge experimented with primitive systems of classification based on the behavior of inmates. The Boston House of Reformation experimented with inmate self-government as a control technique. But despite public declarations to the contrary, there is ample evidence of the use of **solitary confinement**, whipping, and other physical punishments.

Inmates of the houses of refuge labored in large workshops manufacturing shoes, producing brass nails, or caning chairs. Young woman delinquents often were put to work spinning cotton and doing laundry. It is estimated that income generated from labor sold to outside contractors supplied up to 40% of the operating expenses of the houses of refuge. The chief problem for refuge managers was that economic depressions could dry up the demand for labor, and there was not always sufficient work to keep the inmates occupied. Not only were there complaints that contractors abused children, but also that such employment prepared youngsters for only the most menial work.

Youths were committed to the houses of refuge for indeterminate periods of time until the legal age of majority. Release was generally obtained through an apprenticeship by the youths to some form of service. The system was akin to the binding-out practices of earlier times. Males typically were apprenticed on farms, on whaling boats, or in the merchant marine. Young women usually were placed into domestic service. Only rarely was a house-of-refuge child placed in a skilled trade. Apprenticeship decisions often were made to ensure that the child would not be reunited with his or her family because this was presumed to be the root cause of the child's problems. As a result, there are many accounts of siblings and parents vainly attempting to locate their lost relatives.

The founders of the houses of refuge were quick to declare their own efforts successful. Prominent visitors to the institutions, such as Alexis de Tocqueville and Dorothea Dix, echoed the praise of the founders. Managers of the refuges produced glowing reports attesting to the positive results of the houses. Sharp disagreements over the severity of discipline required led to the replacement of directors who were perceived as too permissive. Elijah Devoe (1848), a house of refuge assistant superintendent, wrote poignantly of the cruelties and injustices in these institutions. There are accounts of violence within the institutions as well. Robert Mennel (1973) estimates that approximately 40% of the children escaped either from the institutions or from their apprenticeship placements. The problems that plagued the houses of refuge did not dampen the enthusiasm of the philanthropists, who assumed that the reformation process was a difficult and tenuous business at best.

Public relations efforts proclaiming the success of the houses of refuge helped lead to a rapid proliferation of similar institutions (Rothman, 1971). While special institutions for delinquent and destitute youth increased in numbers, the public perceived that delinquency was continuing to rise and become more serious. The founders of the houses of refuge argued that the solution to the delinquency problem lay in the perfection of better methods to deal with incarcerated children. Most of the literature of this period assumes the necessity of institutionalized treatment for children. The debates centered on whether to implement changes in architecture or in the institutional routines. Advocates of institutionalized care of delinquent and dependent youths continued to play the dominant role in formulating social policy for the next century.

THE GROWTH OF INSTITUTIONALIZATION AND THE CHILD SAVERS (1850–1890)

In the second half of the 19th century, a group of reformers known as the Child Savers instituted new measures to prevent juvenile delinquency (Hawes, 1971; Mennel, 1973; Platt, 1968). Reformers including Lewis Pease, Samuel Gridley Howe, and Charles Loring Brace founded societies to save children from depraved and criminal lives. They created the Five Points Mission (1850), the Children's Aid Society (1853), and the New York Juvenile Asylum (1851). The ideology of this group of reformers differed from that of the founders of the houses of refuge only in that this group was more optimistic about the possibilities of reforming youths. Centers were established in urban areas to distribute food and clothing, provide temporary shelter for homeless youth, and introduce contract systems of shirt manufacture to destitute youth.

The Child Savers criticized the established churches for not doing more about the urban poor. They favored an activist clergy that would attempt to reach the children of the streets. Although this view was somewhat unorthodox, they viewed the urban masses as a potentially dangerous class that could rise up if misery and impoverishment were not alleviated. Charles Loring Brace observed, "Talk of heathen! All the pagans of Golconda would not hold a light to the ragged, cunning, forsaken, godless, keen devilish boys of Leonard Street and the Five Points . . . Our future voters, and President-makers, and citizens! Good Lord deliver us, and help them!" (quoted in Mennel, 1973, p. 34). Brace and his associates knew from firsthand experience in the city missions that the problems of poverty were widespread and growing more serious. Their chief objection to the houses of refuge was that long-term institutionalized care did not reach enough children. Moreover, the Child Savers held the traditional view that family life is superior to institutional routines for generating moral reform.

Brace and his Children's Aid Society believed that delinquency could be solved if vagrant and poor children were gathered up and "placed out" with farm families on the western frontier (O'Connor, 2001). Placing out as a delinquency prevention practice was based on the idealized notion of the U.S. farm family. Such families were supposed to be centers of warmth, compassion, and morality; they were "God's reformatories" for wayward youth. Members of the Children's Aid Society provided food, clothing, and sometimes shelter to street waifs and preached to them about the opportunities awaiting them if they migrated westward. Agents of the Children's Aid Society vigorously urged poor urban youngsters to allow themselves to be placed out with farm families. Many believed that western families provided both a practical and economical resource for reducing juvenile delinquency. The following passage from a Michigan newspaper gives a vivid picture of the placing out process:

> Our village has been astir for a few days. Saturday afternoon, Mr. C. C. Tracy arrived with a party of children from the Children's Aid Society in New York. Sabbath day Mr. Tracy spoke day and evening, three times, in different church edifices to crowded and interested audiences. In the evening, the children were present in a body, and sang their "Westward Ho" song. Notice was given that applicants would find unappropriated children at the store of Carder and Ryder, at nine o'clock Monday morning. Before the hour arrived a great crowd assembled, and in two hours *every child was disposed of,* and more were wanted.
>
> We *Wolverines* will never forget Mr. Tracy's visit. It cost us some tears of sympathy, some dollars, and some smiles. We wish him a safe return to Gotham, a speedy one to us with the new company of destitute children, for whom good homes are even now prepared. (Mennel, 1973, p. 39)

Contrary to the benevolent image projected by this news story, there is ample evidence that the children were obliged to work hard for their keep and were rarely accepted as members of the family. The Boston Children's Aid Society purchased a home in 1864, which was used to help adjust street youth to their new life in the West. The children were introduced to farming skills and taught manners that might be expected of them in their new homes.

Another prevention experiment during the middle part of the 19th century was the result of the work of a Boston shoemaker, John Augustus. In 1841, Augustus began to put up **bail** for men charged with drunkenness, although he had no official connection with the court. Soon after, he extended his services to young people. Augustus supervised the youngsters while they were out on bail, provided clothing and shelter, was sometimes able to find them jobs, and often paid court costs to keep them out of jail. This early **probation** system was later instituted by local child-saving groups, who would find placements for the children. By 1869 Massachusetts had a system by which agents of the Board of State Charities took charge of delinquents before they appeared in court. The youths often were released on probation, subject to good behavior in the future.

These noninstitutional prevention methods were challenged by those who felt an initial period of confinement was important before children were placed out. Critics also argued that the Children's Aid Societies neither followed up on their clients nor administered more stringent discipline to those who needed it. One critic phrased it this way:

> The "vagabond boy" is like a blade of corn, coming up side by side with a thistle. You may transplant both together in fertile soil, but you will have the thistle still. . . . I would have you pluck out the vagabond first, and then let the boy be thus provided with "a home," and not before. (Mennel, 1973, p. 46)

Many Midwesterners were unsettled by the stream of "criminal children" flowing into their midst. Brace and his colleagues were accused of poisoning the West with the dregs of urban life. To combat charges that urban youths were responsible for the rising crime in the West, Brace conducted a survey of western prisons and almshouses to show that few of his children had gotten into further trouble in the West.

Resistance continued to grow against the efforts of the Children's Aid Societies. Brace, holding that asylum interests were behind the opposition, maintained that the longer a child remains in an asylum, the less likely he will reform. (The debate over the advantages and disadvantages of institutionalized care of delinquent youth continues to the present day.) Brace continued to be an active proponent of the placing out system. He appeared before early conventions of reform school managers to present his views and debate the opposition. As the struggle continued over an ideology to guide prevention efforts, the problem of delinquency continued to grow. During the 19th century, poverty, industrialization, and immigration, as well as the Civil War, helped swell the ranks of the "dangerous classes."[9]

Midway through the 19th century, state and municipal governments began taking over the administration of institutions for juvenile delinquents. Early efforts had been supported by private philanthropic groups with some state support. But the growing fear of class strife, coupled with increasing delinquency, demanded a more centralized administration. Many of the newer institutions were termed *reform schools* to imply a strong emphasis on formal schooling. In 1876, of the 51 refuges or reform schools in the United States, nearly three quarters were operated by state or local governments. By 1890, almost every state outside the South had a reform school, and many jurisdictions had separate facilities for male and female delinquents. These institutions varied considerably in their admissions criteria, their sources of referral, and the character of their inmates. Most of the children were sentenced to remain in reform schools until they reached the age of majority (18 years for girls and 21 for boys) or until they were reformed. The length of confinement, as well as the decision to transfer unmanageable youths to adult penitentiaries, was left to the discretion of reform school officials.

Partially in response to attacks by Brace and his followers, many institutions implemented a cottage or family system between 1857 and 1860. The cottage system involved dividing the youths into units of 40 or fewer, each with its own cottage and schedule. Although work was sometimes performed within the cottages, the use of large congregate workshops continued. The model for the system was derived from the practice of European correctional officials. There is evidence from this period of the development of a self-conscious attempt to refine techniques to mold, reshape, and reform wayward youths (Hawes, 1971).

During this period, a movement was initiated to locate institutions in rural areas because it was felt that agricultural labor would facilitate reformative efforts. As a result, several urban houses of refuge were relocated in rural settings. Many rural institutions used the cottage system, as it was well suited to agricultural production. In addition, the cottage system gave managers the opportunity to segregate children according to age, sex, race, school achievement, or "hardness." Critics of the institutions, such as Mary Carpenter, pointed out that most of the presumed benefits of rural settings were artificial and that the vast majority of youths who spent time in these reform schools ultimately returned to crowded urban areas.

The Civil War deeply affected institutions for delinquent youth. Whereas prisons and county jails witnessed declines in population, the war brought even more youths into reform schools. Institutions were strained well beyond their capacities. Some historians believe that the participation of youths in the draft riots in northern cities produced an increase in incarcerated youths. Reform schools often released older youngsters to military service to make room for additional children. Due to the high inflation rates of the war, the amount of state funds available for institutional upkeep steadily declined. Many institutions were forced to resort to the contract labor system to increase reform school revenues to meet operating expenses during the war and in the postwar period.

Voices were raised in protest over the expansion of contract labor in juvenile institutions. Some charged that harnessing the labor of inmates, rather than the reformation of youthful delinquents, had become the raison d'être of these institutions. There were growing rumors of cruel and vicious exploitation of youth by work supervisors. An 1871 New York Commission on Prison Labor, headed by Enoch Wines, found that refuge boys were paid 30 cents per day for labor that would receive 4 dollars a day on the outside. In the Philadelphia House of Refuge, boys were paid 25 cents a day and were sent elsewhere if they failed to meet production quotas. Economic depressions throughout the 1870s increased pressure to end the contract system. Workingmen's associations protested against the contract system because prison and reform school laborers created unfair competition. Organized workers claimed that refuge managers were making huge profits from the labor of their wards:

> From the institutional point of view, protests of workingmen had the more serious result of demythologizing the workshop routine. No longer was it believable for reform school officials to portray the ritual as primarily a beneficial aid in inculcating industrious habits or shaping youth for "usefulness." The violence and exploitation characteristic of reform school workshops gave the lie to this allegation. The havoc may have been no greater than that which occasionally wracked the early houses of refuge, but the association of conflict and the contract system in the minds of victims and outside labor interests made it now seem intolerable. (Mennel, 1973, p. 61)

The public became aware of stabbings, fighting, arson, and attacks upon staff of these institutions. All signs pointed toward a decline of authority within the institutions. The economic troubles of the reform schools continued to worsen. Additional controversy was generated by organized Catholic groups, who objected to Protestant control of juvenile institutions housing a majority of Catholics. This crisis in the juvenile institutions led to a series of investigations into reform school operations.[10] The authors of these reports proposed reforms to maximize efficiency of operation and increase government control over the functioning of institutions in their jurisdictions. One major result of these investigative efforts was the formation of Boards of State Charity. Members of these boards were appointed to inspect reform schools and make recommendations for improvements but were to avoid the evils of the patronage system. Board members, who were described as "gentlemen of public spirit and sufficient leisure," uncovered horrid institutional conditions and made efforts to transfer youngsters to more decent facilities. Men such as Frederick Wines, Franklin Sanborn, Hastings Hart, and William Pryor Letchworth were among the pioneers of this reform effort (Mennel, 1973, p. 61).

Although it was hoped that the newly formed boards would find ways to reduce the proliferation of juvenile institutions, such facilities continued to grow, as did the number of wayward youths. These late-19th-century reformers looked toward the emerging scientific disciplines for solutions to the problems of delinquency and poverty. They also developed a system to discriminate among delinquents, so that "hardened offenders" would be sent to special institutions such as the Elmira Reformatory. It was generally recognized that new methods would have to be developed to restore order within the reform schools and to make some impact upon delinquency.

Juvenile institutions in the South and the far West developed much later than those in the North or the East, but did so essentially along the same lines. One reason for this was that delinquency was primarily a city problem, and the South and far West were less urbanized (Mintz, 2004). Another reason was that in the South, black youths received radically different treatment from whites. Whereas there was toleration for the misdeeds of white youth, black children were controlled under the disciplinary systems of slavery. Even after Emancipation, the racism of southern whites prevented them from treating black children as fully human and worth reforming (Ward, 2012). The Civil War destroyed the prison system of the South. After the war, southern whites used the notorious Black Codes and often trumped up criminal charges to arrest thousands of impoverished former slaves, placing them into a legally justified forced labor system. Blacks were leased out on contract to railroad companies, mining interests, and manufacturers. Although many of these convicts were children, no special provisions were made because of age. Conditions under the southern convict lease system were miserable and rivaled the worst cruelties of slavery. Little in the way of specialized care for delinquent youth was accomplished in the South until well into the 20th century. The convict lease system was eventually replaced by county road gangs and prison farms, characterized by grossly inhumane conditions of confinement. These were systems of vicious exploitation of labor and savage racism (McKelvey, 1972).

JUVENILE DELINQUENCY AND THE PROGRESSIVE ERA

The period from 1880 to 1920, often referred to by historians as the Progressive Era, was a time of major social structural change in the United States. The nation was in the process of becoming increasingly urbanized, and unprecedented numbers of European immigrants were migrating to cities in the Northeast. The United States was becoming an imperialist power and was establishing worldwide military and economic relationships. Wealth was becoming concentrated in the hands of a few individuals who sought to dominate U.S. economic life. Labor violence was on the rise, and the country was in the grip of a racial hysteria affecting all peoples of color. The tremendous technological developments of the time reduced the need for labor (Weinstein, 1968; Williams, 1973).

During the Progressive Era, those in positions of economic power feared that the urban masses would destroy the world they had built. Internal struggles developing among the wealthy heightened the tension. From all sectors came demands that new action be taken to preserve social order and to protect private property and racial privilege (Gossett, 1963). Up to this time, those in positions of authority had assumed a laissez-faire stance, fearing that government intervention might extend to economic matters. Although there was general agreement on the need for law enforcement to maintain social order, there was profound skepticism about attempts to alleviate miserable social conditions or reform deviant individuals. Some suggested that if society consisted of a natural selection process in which the fittest would survive, then efforts to extend the life chances of the poor or "racially inferior" ran counter to the logic of nature.

Others during this era doubted the wisdom of a laissez-faire policy and stressed that the threat of revolution and social disorder demanded scientific and rational methods to restore social order. The times demanded reform, and before the Progressive Era ended, much of the modern welfare state and the criminal justice system were constructed. Out of the turmoil of this age came such innovations as widespread use of the indeterminate sentence, the public defender movement, the beginning of efforts to professionalize the police, extensive use of **parole**, the rise of mental and IQ testing, scientific study of crime, and ultimately the juvenile court (Chavez-Garcia, 2012).

Within correctional institutions at this time, there was optimism that more effective methods would be found to rehabilitate offenders. One innovation was to institute physical exercise training, along with special massage and nutritional regimens. Some believed that neglect of the body had a connection with delinquency and crime. Those who emphasized the importance of discipline in reform efforts pressed for the introduction of military drill within reform schools. There is no evidence that either of these treatment efforts had a reformative effect upon inmates, but it is easy to understand why programs designed to keep inmates busy and under strict discipline would be popular at a time of violence and disorder within prisons and reform schools. As institutions faced continual financial difficulties, the contract labor system came under increasing attack. Criticism of reform schools resulted in laws in some states to exclude children under the age of 12 from admission to reform schools. Several states abolished the contract labor system, and efforts were made to guarantee freedom of worship among inmates of institutions. Once again, pleas were made for community efforts to reduce delinquency rather than society relying solely upon reform schools as a prevention strategy. The arguments put forth were reminiscent of those of Charles Loring Brace and the Child Savers. For example, Homer Folks, president of the Children's Aid Society of Pennsylvania, articulated these five major problems of reform schools in 1891:

1. The temptation it offers to parents and guardians to throw off their most sacred responsibilities.

2. The contaminating influence of association.

3. The enduring stigma . . . of having been committed . . . .

4. . . . renders impossible the study and treatment of each child as an individual.

5. The great dissimilarity between life in an institution and life outside. (Mennel, 1973, p. 111)

One response was to promote the model of inmate self-government within the institution's walls. One such institution, the George Junior Republic, developed an elaborate system of inmate government in 1893, in which the institution became a microcosm of the outside world. Self-government was viewed as an effective control technique because youths became enmeshed in the development and enforcement of rules, while guidelines for proper behavior continued to be set by the institutional staff. The inmates were free to construct a democracy so long as it conformed to the wishes of the oligarchic staff (Hawes, 1971).

The populist governments of several southern states built reform schools, partly due to their opposition to the convict lease system. But these institutions too were infused with the ethos of the **Jim Crow laws**, which attempted to permanently legislate an inferior role for black Americans in southern society. One observer described the reform school of Arkansas as a place "where White boys might be taught some useful occupation and the negro boys compelled to work and support the institution while it is being done" (Mennel, 1973, p. 12). Black citizens, obviously displeased with discrimination within southern reform schools, proposed that separate institutions for black children should be administered by the black community (Ward, 2012). A few such institutions were established, but the majority of black children continued to be sent to jail or to be the victims of lynch mobs.

Growing doubt about the success of reform schools in reducing delinquency led some to question the wisdom of applying an unlimited *parens patriae* doctrine to youth. In legal cases, such as *The People v. Turner* (1870), *State v. Ray* (1886), and *Ex parte Becknell* (1897), judges questioned the quasi-penal character of juvenile institutions and wondered whether there ought not to be some procedural safeguards for children entering court on delinquency charges.

The state of Illinois, which eventually became the first state to establish a juvenile court law, had almost no institutions for the care of juveniles. Most early institutions in Illinois had been destroyed in fires, and those that remained were regarded as essentially prisons for children. Illinois attempted a privately financed system of institutional care, but this also failed. As a result, progressive reformers in Chicago complained of large numbers of children languishing in the county jail and pointed out that children sometimes received undue leniency due to a lack of adequate facilities.

A new wave of Child Savers emerged, attempting to provide Chicago and the state of Illinois with a functioning system for handling wayward youth.[11] These reformers, members of the more wealthy and influential Chicago families, were spiritual heirs of Charles Loring Brace in that they, too, feared that social unrest could destroy their authority. But through their approach, they hoped to alleviate some of the suffering of the impoverished and ultimately win the loyalty of the poor. Reformers such as Julia Lathrop, Jane Addams, and Lucy Flower mobilized the Chicago Women's Club on behalf of juvenile justice reform. Other philanthropic groups, aligning with the powerful Chicago Bar Association, helped promote a campaign leading to the eventual drafting of the first juvenile court law in the United States. Although previous efforts had been made in Massachusetts and Pennsylvania to initiate separate trials for juveniles, the Illinois law is generally regarded as the first comprehensive child welfare legislation in this country.

The Illinois law, passed in 1899, established a children's court that would hear cases of delinquent, dependent, and neglected children. The *parens patriae* philosophy, which had imbued the reform schools, now extended to the entire court process. The definition of delinquency was broad, so that a child would be adjudged delinquent if he or she violated any state law or any city or village ordinance. In addition, the court was given jurisdiction in cases of incorrigibility, truancy, and lack of proper parental supervision. The court had authority to institutionalize children, send them to orphanages or foster homes, or place them on probation. The law provided for unpaid probation officers, who would assist the judges and supervise youngsters. In addition, the law placed the institutions for dependent youth under the authority of the State Board of Charities and regulated the activities of agencies sending delinquent youth from the East into Illinois.

The juvenile court idea spread so rapidly that within 10 years of the passage of the Illinois law, 10 states had established children's courts. By 1912, 22 states had juvenile court laws, and by 1925, all but two states had established specialized courts for children. Progressive reformers proclaimed the establishment of the juvenile court as the most significant reform of this period. The reformers celebrated what they believed to be a new age in the treatment of destitute and **delinquent children**. In *Commonwealth v. Fisher* (1905), the Pennsylvania Supreme Court defended the juvenile court ideal in terms reminiscent of the court opinion in the *Crouse* case of 1838:

> To save a child from becoming a criminal, or continuing in a career of crime, to end in mature years in public punishment and disgrace, the legislatures surely may provide for the salvation of such a child, if its parents or guardians be unwilling or unable to do so, by bringing it into one of the courts of the state without any process at all, for the purpose of subjecting it to the state's guardianship and protection.

Critics, pointing to the large number of children who remained in jails and detention homes for long periods, expressed doubt that the court would achieve its goal. Some judges, including the famous Judge Ben Lindsey of Denver, decried the seemingly unlimited discretion of the court. With so much diversity among jurisdictions in the United States, it is difficult to describe the functioning of a typical court. As the volume of cases in the urban areas soon overwhelmed existing court resources, judges became unable to give the close personal attention to each case advocated by the reformers. As little as 10 minutes were devoted to each case as court calendars became increasingly crowded. Similarly, as **caseloads** soared, the quality of probationary supervision deteriorated and became perfunctory.

It is important to view the emergence of the juvenile court in the context of changes taking place in U.S. society at that time. Juvenile court drew support from a combination of optimistic social theorists, sincere social reformers, and the wealthy, who felt a need for social control. The juvenile court movement has been viewed as an attempt to stifle legal rights of children by creating a new adjudicatory process based on principles of equity law. This view misses the experimental spirit of the Progressive Era by assuming a purely conservative motivation on the part of the reformers.

Although most reformers of the period understood the relationship between poverty and delinquency, they responded with vastly different solutions. Some reformers supported large-scale experimentation with new social arrangements, such as the Cincinnati Social Unit Experiment, an early forerunner of the community organization strategy of the War on Poverty of the 1960s (Shaffer, 1971). Other reformers looked to the emerging social science disciplines to provide a rational basis for managing social order. During the Progressive Era, there was growth in the profession of social work, whose members dealt directly with the poor.[12] Progressive reformers conducted social surveys to measure the amount of poverty, crime, and juvenile dependency in their communities. They supported social experiments to develop new behavior patterns among the lower classes to help them adjust to the emerging corporate economy. The development of mental testing became crucial in defining access to the channels of social mobility and for demonstrating, to the satisfaction of the white ruling class, their own racial superiority. Moreover, biological explanations of individual and social pathology rationalized the rise in crime and social disorder without questioning the justice or rationality of existing social arrangements (Chavez-Garcia, 2012).

The thrust of Progressive Era reforms was to found a more perfect control system to restore social stability while guaranteeing the continued hegemony of those with wealth and privilege.

Reforms such as the juvenile court are ideologically significant because they preserved the notion that social problems (in this case, delinquency, dependency, and neglect) could be dealt with on a case-by-case basis rather than through broad-based efforts to redistribute wealth and power throughout society. The chief dilemma for advocates of the juvenile court was to develop an apparently apolitical or neutral system while preserving differential treatment for various groups of children. The juvenile court at first lacked a core of functionaries who could supply the rationale for individualized care for wayward youth, but soon these needs were answered by the emergence of psychiatry, psychology, and criminology, as well as by the expanding profession of social work.

THE CHILD GUIDANCE CLINIC MOVEMENT

In 1907, Illinois modified its juvenile court law to provide for paid probation officers, and the Chicago Juvenile Court moved into new facilities with expanded detention space. The Juvenile Protective League, founded by women active in establishing the first juvenile court law, was intended to stimulate the study of the conditions leading to delinquency. The members of the Juvenile Protective League were especially troubled that large numbers of wayward youth repeatedly returned to juvenile court. Jane Addams, a major figure in U.S. philanthropy and social thought, observed, "At last it was apparent that many of the children were psychopathic cases and they and other borderline cases needed more skilled care than the most devoted probation officer could give them" (Hawes, 1971, p. 244).

But the new court facilities did provide an opportunity to examine and study all children coming into the court. The Juvenile Protective League promised to oversee this study of delinquency, and Ellen Sturges Dummer donated the necessary money to support the effort. Julia Lathrop was chosen to select a qualified psychologist to head the project. After consulting with William James, she selected one of his former students, William A. Healy. Healy proposed a 4- to 5-year study to compare some 500 juvenile court clients with patients in private practice. The investigation, according to Healy, "would have to involve all possible facts about heredity, environment, antenatal and postnatal history, etc." (Hawes, 1971, p. 250).

In 1909, the Juvenile Protective League established the Juvenile Psychopathic Institute, with Healy as its first director and Julia Lathrop, Jane Addams, and Judge Julian W. Mack on the executive committee.[13] The group, in its opening statement, expressed its plans to "undertake . . . an inquiry into the health of delinquent children in order to ascertain as far as possible in what degrees delinquency is caused or influenced by mental or physical defect or abnormality and with the purpose of suggesting and applying remedies in individual cases whenever practicable as a concurrent part of the inquiry" (Hawes, 1971, pp. 250–251).

Jane Addams added her concern that the study investigate the conditions in which the children lived as well as the mental and physical history of their ancestors.

Healy held an MD degree from the University of Chicago and had served as a physician at the Wisconsin State Hospital. He had taught university classes in neurology, mental illness, and gynecology; had studied at the great scientific centers of Europe; and was familiar with the work of Sigmund Freud and his disciples. The major tenet of Healy's scientific credo was that the individual was the most important unit for study. Healy argued that the individualization of treatment depended upon scientific study of individual delinquents.

Healy and his associates published *The Individual Delinquent: A Textbook of Diagnosis and Prognosis for All Concerned in Understanding Offenders* in 1915. This book, based on a study of

CASE STUDY: THE HISTORICAL LEGACY OF JUVENILE JUSTICE

In 1838, Mary Ann Crouse, who was under 18 years old, was committed by a local magistrate to the Philadelphia House of Refuge because her mother claimed that she could not control the young woman due to her "vicious behavior." The placement was done without any hearing. Mary Ann's father filed a legal appeal saying that his property rights to his daughter had been violated without cause. The Pennsylvania Supreme Court denied the father's claim and established the powerful legal precedent of *parens patriae*, saying that the state had an obligation to assume the obligations to protect the child if her parents could not properly control her behavior. The court argued that to release her from custody would be a greater act of cruelty. This important decision became a core judicial pillar of the American juvenile court.

1,000 cases of repeat juvenile offenders, was intended as a practical handbook. The methodology involved a study of each offender from social, medical, and psychological viewpoints. Healy even did anthropometric measurements, suggested by Cesare Lombroso and his followers, although Healy doubted that delinquents formed a distinctive physical type.[14] However, Healy never was able to locate a limited set of causes for delinquency through empirical observation. He stressed the wide range of potential causes of delinquency, including the influence of bad companions, the love of adventure, early sexual experiences, and mental conflicts. At this stage, Healy adopted an eclectic explanation of delinquency: "Our main conclusion is that every case will always need study by itself. When it comes to arraying data for the purpose of generalization about relative values of causative factors we experience difficulty" (Mennel, 1973, p. 165). Despite exhaustive research, Healy and his associates could not find distinctive mental or physical traits to delineate delinquents from nondelinquents.

Later, in 1917, Healy advanced his theory of delinquency in *Mental Conflicts and Misconduct*. In this work, Healy stressed that although individuals may experience internal motivation toward misbehavior, this usually results in their merely feeling some anxiety. When mental conflict becomes more acute, the child may respond by engaging in misconduct. These ideas were heavily influenced by the work of Adolf Meyer, whose interpretation of Freud had a major influence on U.S. psychiatry. Healy agreed with Meyer that the family was a crucial factor in delinquency: "The basis for much prevention of mental conflict is to be found in close comfortable relations between parents and children" (Hawes, 1971, p. 255). Healy's emphasis on the family was well received by those in the delinquency prevention field who had traditionally viewed the family as God's reformatory.

The significance of Healy's work cannot be overemphasized, as it provided an ideological rationale to defend the juvenile court. Healy's work gave legitimacy to the flexible and discretionary operations of the court. Although some used Healy's emphasis on the individual to minimize the importance of social and economic injustice, there is evidence that Healy understood that delinquency was rooted in the nature of the social structure:

> If the roots of crime lie far back in the foundations of our social order, it may be that only a radical change can bring any large measure of cure. Less unjust social and economic conditions may be the only way out, and until a better social order exists, crime will probably continue to flourish and society continue to pay the price. (Healy, Bronner, & Shimberg, 1935, p. 211)

Healy's work also gave support to the concept of professionalism in delinquency prevention. Because juvenile delinquency was viewed as a complex problem with many possible causes, this rationale was used to explain the increased reliance on experts. In the process, the juvenile court became insulated from critical scrutiny by its clients and the community. If actions taken by the court did not appear valid to the layman, this was because of a higher logic, known only to the experts, which explained that course of action. Moreover, the failure of a specific treatment program often was attributed to the limits of scientific knowledge or to the failure of the court to follow scientific principles in its dispositions.

After his work in Chicago, Healy went to the Judge Harvey Baker Foundation in Boston to continue his research, where he began actual treatment of youths. Healy became a proselytizer for the child guidance clinic idea. Working with the Commonwealth Fund and the National Committee for Mental Hygiene, Healy aided the development of child guidance clinics across the nation. These efforts were so successful that by 1931, 232 such clinics were in operation. There is even a report of a traveling child guidance clinic that visited rural communities in the West to examine children. The child guidance clinic movement became an important part of a broader campaign to provide mental hygiene services to all young people. The clinics initially were set up in connection with local juvenile courts, but later some of them became affiliated with hospitals and other community agencies.

In Sheldon and Eleanor Glueck's (1934) classic delinquency research, they evaluated the success of Healy's Boston clinic. In *One Thousand Juvenile Delinquents: Their Treatment by Court and Clinic*, the Gluecks found high rates of recidivism among children treated at the clinic. Healy, though deeply disappointed by the results, continued his efforts. The Gluecks continued, in a series of longitudinal studies, to search for the causes of delinquency and crime.[15] Like Healy, they maintained a focus on the individual, and they increased efforts to discover the factors behind repeated delinquency. The work of the Gluecks reflected a less optimistic attitude about the potential for treatment and rehabilitation than that found in Healy's work. They emphasized the importance of the family, often ignoring the impact of broader social and economic factors. It is ironic that the thrust of delinquency theories in the 1930s should be toward individual and family conflicts. As 20% of the American people were unemployed, the effects of the Great Depression of the 1930s must have been apparent to the delinquents and their families, if not to the good doctors who studied them with such scientific rigor.

THE CHICAGO AREA PROJECT

The Chicago Area Project of the early 1930s is generally considered the progenitor of large-scale, planned, community-based efforts with delinquent youth. The project differed from the dominant approaches of the time, which relied on institutional care and psychological explanations for delinquent behavior. The Chicago Area Project, conceived by University of Chicago sociologist Clifford Shaw, was an attempt to implement a sociological theory of delinquency in the delivery of preventive services. The theoretical heritage of the project is found in such works as *The Jack-Roller, a Delinquent Boy's Own Story* (1930), *Brothers in Crime* (1938), and *Juvenile Delinquency and Urban Areas* (1942), all written by Shaw and his associates. They attributed variations in delinquency rates to demographic or socioeconomic conditions in different areas of cities. This environmental approach assumed that delinquency was symptomatic of social disorganization. The adjustment problems of recent immigrants, together with other problems of urban life, strained the influence on adolescents of traditional social control agencies such as family, church, and community. Delinquency was viewed as a problem of the modern city, which was characterized by the breakdown of spontaneous or natural forces of social control. Shaw contended that the rapid social change that migrant rural youths are subjected to when entering the city promotes alienation from accepted modes of behavior:

> When growing boys are alienated from institutions of their parents and are confronted with a vital tradition of delinquency among their peers, they engage in delinquent activity as part of their groping for a place in the only social groups available to them. (Kobrin, 1970, p. 579)

The Chicago Area Project thus viewed delinquency as "a reversible accident of the person's social experience" (Kobrin, 1970).

The project employed several basic operating assumptions. The first was that the delinquent is involved in a web of daily relationships. As a result, the project staff attempted to mobilize adults in the community, hoping to foster indigenous neighborhood leadership to carry out the programs with delinquent youth. The second assumption was that people participate only if they have meaningful roles; therefore, the staff attempted to share decision making with neighborhood residents. To maximize community participation, staff members had to resist the urge to direct the programs themselves. The final premise of the Chicago Area Project was that within a given community there are people who, when given proper training and guidance, can organize and administer local welfare programs. A worker from within the community who has knowledge of local customs and can communicate easily with local residents is more effective in dealing with delinquency problems. The project staff believed that placing community residents in responsible positions would demonstrate the staff's confidence in the ability of residents to solve their own problems.

The Chicago Area Project was overseen by a board of directors responsible for raising and distributing funds for research and community programs. In several years, 12 community committees developed in Chicago as "independent, self-governing, citizens' groups, operating under their own names and charters" (Sorrento, quoted in Sechrest, 1970, p. 6). The neighborhood groups were aided by the board in obtaining grants to match local funds. Personnel from the Institute for Juvenile Research at the University of Chicago served as consultants to local groups. The various autonomous groups pursued such activities as the creation of recreation programs or community-improvement campaigns for schools, traffic safety, sanitation, and law enforcement. There were also programs aimed directly at delinquent youth, such as visitation privileges for incarcerated children, work with delinquent gangs, and volunteer assistance in parole and probation.

Most observers have concluded that the Chicago Area Project succeeded in fostering local community organizations to attack problems related to delinquency (Kobrin, 1970; Shaw & McKay, 1942). Evidence also shows that delinquency rates decreased slightly in areas affected by the project, but these results are not conclusive. Shaw explained the difficulty of measuring the impact of the project as follows:

> Conclusive statistical proof to sustain any conclusion regarding the effectiveness of this work in reducing the volume of delinquency is difficult to secure for many reasons. Trends in rates for delinquents for small areas are affected by variations in the definition of what constitutes delinquent behavior, changes in the composition of the population, and changes in the administrative procedures of law enforcement agencies. (Witmer & Tufts, 1954, p. 16)

The Illinois State Division of Youth Services took over all 35 staff positions of the Chicago Area Project in 1957. It appears that this vibrant and successful program was quickly transformed into "a rather staid, bureaucratic organization seeking to accommodate itself to the larger social structure, that is, to work on behalf of agencies who came into the community rather than for itself or for community residents" (Sechrest, 1970, p. 15).

The Chicago Area Project, with its grounding in sociological theory and its focus on citizen involvement, contrasts sharply with other delinquency prevention efforts of the 1930s. Its focus on prevention in the community raised questions about the continued expansion of institutions for delinquent youth. Although some attributed support of the project to the personal dynamism of Clifford Shaw, this ignores the basic material and ideological motivation behind it. It would be equally shortsighted to conclude that child saving would not have occurred without Charles Loring Brace or that the child guidance clinic movement resulted solely from the labors of William Healy. Certainly Shaw was an important advocate of the Chicago Area Project approach, and his books influenced professionals in the field, but the growth of the project was also a product of the times.

Because no detailed history exists of the founding and operation of the project, we can only speculate about the forces that shaped its development. We do know that Chicago at that time was caught in the most serious economic depression in the nation's history. Tens of thousands of people were unemployed, especially immigrants and blacks. During this period, a growing radicalization among impoverished groups resulted in urban riots (Cloward & Piven, 1971). The primary response by those in positions of power was to expand and centralize charity and welfare systems. In addition, there was considerable experimentation with new methods of delivering relief services to the needy. No doubt, Chicago's wealthy looked favorably upon programs such as the Chicago Area Project, which promised to alleviate some of the problems of the poor without requiring a redistribution of wealth or power. Both the prestige of the University of Chicago and the close supervision promised by Shaw and his associates helped assuage the wealthy and the powerful. Shaw and his associates did not advocate fundamental social change, and project personnel were advised to avoid leading communities toward changes perceived as too radical (Alinsky, 1946). Communities were encouraged to work within the system and to organize around issues at a neighborhood level. Project participants rarely questioned the relationship of urban conditions to the political and economic superstructure of the city.

Later interpreters of the Chicago Area Project did not seem to recognize the potentially radical strategy of community organization within poor neighborhoods. Its immediate legacy was twofold—the use of detached workers, who dealt with gangs outside the agency office, and the idea of using indigenous workers in social control efforts. Although detached workers became a significant part of the delinquency prevention strategy of the next three decades, the use of indigenous personnel received little more than lip service because welfare and juvenile justice agencies hired few urban poor.

The success of the Chicago Area Project depended upon relatively stable and well-organized neighborhoods with committed local leaders. Changes in the urban structure that developed over the next two decades did not fit the Chicago Area Project model. The collapse of southern agriculture and mass migration by rural blacks into the cities of the North and West produced major social structural changes. This movement to the North and West began in the 1920s, decreased somewhat during the Great Depression years, and later accelerated due to the attraction provided by the war industry jobs. During this same period, large numbers of Puerto Ricans settled in New York City and other eastern cities. Although economic opportunity attracted new migrants to the urban centers, there was little satisfaction for their collective dreams. Blacks who left the South to escape the Jim Crow laws soon were confronted by de facto segregation in schools, in the workplace, and in housing. Job prospects were slim for blacks and Puerto Ricans, and both groups were most vulnerable to being fired at the whims of employers. In many respects, racism in the North rivaled that of the South. The new migrants had the added difficulty of adapting their primarily rural experiences to life in large urban centers (Coles, 1967; Handlin, 1959).

Racialized ghettos became places of poverty, disease, and crime. For the more privileged classes, the situation paralleled that of 16th-century European city dwellers who feared the displaced peasantry, or that of Americans at the beginning of the 19th century who feared the Irish immigrants. During this period, riots erupted in East St. Louis, Detroit, Harlem, and Los Angeles. To upper-class observers, these new communities of poor black and brown peoples were disorganized collections of criminals and deviants. Racism prevented white observers from recognizing the vital community traditions or the family stability that persisted despite desperate economic conditions. Moreover, the label *disorganized communities* could be used ideologically to mask the involvement of wealthy whites in the creation of racial ghettos (Ryan, 1971).

A liberal social theory was developing that, though benign on the surface, actually blamed the victims for the conditions in which they were caught. Attention was focused upon deviant aspects of community life, ascribing a culture of poverty and violence to inner-city residents and advocating remedial work with individuals and groups to solve so-called problems of adjustment. The following quote from the National Commission on the Causes and Prevention of Violence (1969) is illustrative of this posture:

> The cultural experience which Negroes brought with them from segregation and discrimination in the rural South was of less utility in the process of adaption to urban life than was the cultural experience of many European immigrants. The net effect of these differences is that urban slums have tended to become ghetto slums from which escape has been increasingly difficult. (p. 30)

Delinquency theorists suggested that lower-class communities were becoming more disorganized because they were not characterized by the stronger ties of older ethnic communities:

> Slum neighborhoods appear to us to be undergoing progressive disintegration. The old structures, which provided social control and avenues of social ascent, are breaking down. Legitimate but functional substitutes for these traditional structures must be developed if we are to stem the trend towards violence and retreatism among adolescents in urban slums. (Cloward & Piven, 1971, p. 211)

Irving Spergel (1996), leading authority on juvenile gangs, suggests that social work agencies made little use of indigenous workers after World War II because delinquency had become more aggressive and violent. Welfare and criminal justice officials argued that only agencies with sound funding and strong leadership could mobilize the necessary resources to deal with the increased incidence and severity of youth crime.

The movement toward more agency involvement brought with it a distinctly privileged-class orientation toward delinquency prevention. Social service agencies were preeminently the instruments of those with sufficient wealth and power to enforce their beliefs. The agencies were equipped to redirect, rehabilitate, and in some cases control those who seemed most threatening to the status quo. Workers for these agencies helped to perpetuate a conception of proper behavior for the poor consistent with their expected social role. For example, the poor were told to defer gratification and save for the future, but the rich often were conspicuous consumers. Whereas poor women were expected to stay at home and raise their families, the same conduct was not uniformly applied to wealthy women. The well to do provided substantial funding for private social service agencies and often became members of the boards that defined policies for agencies in inner-city neighborhoods. The criteria for staffing these agencies during the two decades following World War II included academic degrees and special training that were not made available to the poor or to people of color.

Social agencies, ideologically rooted and controlled outside poor urban neighborhoods, were often pressured to respond to "serious" delinquency problems. During this period, the fighting gang, which symbolized organized urban violence, received the major share of delinquency prevention efforts. Most agencies, emphasizing psychoanalytic or group dynamic approaches to delinquency, located the origin of social disruption in the psychopathology of individuals and small groups. The consequence of this orientation was that special youth workers were assigned to troublesome gangs in an attempt to redirect the members toward more conventional conduct. Little effort was made to develop local leadership or to confront the issues of racism and poverty.

Detached worker programs emphasized treatment by individual workers freed from the agency office base and operating in neighborhood settings. These programs, with several variations, followed a basic therapeutic model. Workers initially entered gang territories, taking pains to make their entrance as inconspicuous as possible. The first contacts were made at natural meeting places in the community such as pool rooms, candy stores, or street corners:

> Accordingly, the popular image of the detached worker is a young man in informal clothing, standing on a street corner near a food stand, chatting with a half dozen rough, ill-groomed, slouching teenagers. His posture is relaxed, his countenance earnest, and he is listening to the boys through a haze of cigarette smoke. (Klein, 1969, p. 143)

The worker gradually introduced himself to the gang members. He made attempts to get jobs for them or arranged recreational activities, while at the same time persuading the members to give up their illegal activities. Manuals for detached workers explained that the approach would work because gang members had never before encountered sympathetic, nonpunitive adults who were not trying to manipulate them for dishonest purposes. A typical report states, "Their world (as they saw it) did not contain any giving, accepting people—only authorities, suckers and hoodlums like themselves" (Crawford, Malamud, & Dumpson, 1970, p. 630). This particular account even suggests that some boys were willing to accept the worker as an "idealized father." The worker was expected to influence the overall direction of the gang, but if that effort failed, he was to foment trouble among members and incite disputes over leadership. Information that the workers gathered under promises of confidentiality was often shared with police gang-control officers. Thus, despite their surface benevolence, these workers were little more than undercover agents whose ultimate charge was to break up or disrupt groups that were feared by the establishment. These techniques, which focused on black and Latino youth gangs in the 1950s, were similar to those later used with **civil rights** groups and organizations protesting the Vietnam War.

There were many critics of the detached worker programs. Some argued that the workers actually lent status to fighting gangs and thus created more violence. Other critics claimed that the workers often developed emotional attachments to youthful gang members and were manipulated by them (Mattick & Caplan, 1967). Community residents often objected to the presence of detached workers because it was feared they would provide information to downtown social welfare agencies. Although studies of the detached worker programs did not yield positive results, virtually all major delinquency programs from the late 1940s to the 1960s used detached workers in an attempt to reach the fighting gang.

THE MOBILIZATION FOR YOUTH

During the late 1950s, economic and social conditions were becoming more acute in the urban centers of the United States. The economy was becoming sluggish, and unemployment began to rise. Black teenagers experienced especially high unemployment rates, and the discrepancy between white and black income and material conditions grew each year. Technological changes in the economy continually drove more unskilled laborers out of the labor force. Social scientists such as Daniel Moynihan (1969) and Sidney Wilhelm (1970) view this period as the time in which a substantial number of blacks became permanently unemployed. Social control specialists for the privileged class surveyed the problem and sought ways to defuse the social danger of a surplus labor population.

The Ford Foundation was influential during this period in stimulating conservative local officials to adopt more enlightened strategies in dealing with the poor (Marris & Rein, 1967; Moynihan, 1969). Once again, an ideological clash occurred between those favoring scientific and rational government programs and those who feared the growth of the state, demanded balanced government budgets, and opposed liberal programs to improve the quality of life of the poor. The Ford Foundation, through its Grey Area projects, spent large amounts of money in several U.S. cities to foster research and planning of new programs to deal with delinquency and poverty.

The most significant program to develop out of the Grey Area projects was the Mobilization for Youth (MFY), which began in New York City in 1962 after 5 years of planning. It aimed to service a population of 107,000 (approximately one third black and Puerto Rican), living in 67 blocks of New York City's Lower East Side. The unemployment rate of the area was twice that of the city overall, and the delinquency rate was also high. The theoretical perspective of the project was drawn from the work of Richard Cloward and Lloyd Ohlin:

> "a unifying principle of expanding opportunities has worked out as the direct basis for action." This principle was drawn from the concepts outlined by the sociologists Richard Cloward and Lloyd Ohlin in their book *Delinquency and Opportunity*. Drs. Cloward and Ohlin regarded delinquency as the result of the **disparity** perceived by low-income youths between their legitimate aspirations and the opportunities—social, economic, political, education—made available to them by society. If the gap between opportunity and aspiration could be bridged, they believed delinquency could be reduced; that would be the agency's goal. (Weissman, 1969, p. 19)

The MFY project involved five areas—work training, education, group work and community organization, services to individuals and families, and training and personnel—but the core of the mobilization was to organize area residents to realize "the power resources of the community by creating channels through which consumers of social welfare services can define their problems and goals and negotiate on their own behalf" (Brager & Purcell, 1967, p. 247). Local public and private bureaucracies became the targets of mass protests by agency workers and residents. The strategy of MFY assumed that social conflict was necessary in the alleviation of the causes of delinquency. Shortly after MFY became directly involved with struggles over the redistribution of power and resources, New York City officials charged that the organization was "riot-producing, Communist-oriented, left-wing and corrupt" (Weissman, 1969, pp. 25–28). In the ensuing months, the director resigned, funds were limited, and virtually all programs were stopped until after the 1964 **presidential** election. After January 1965, MFY moved away from issues and protests toward more traditional approaches to social programming, such as detached-gang work, job training, and counseling.

Another project, Harlem Youth Opportunities Unlimited (Haryou-Act), which developed in the black community of Harlem in New York City, experienced a similar pattern of development and struggle. The Harlem program was supported by the theory and prestige of psychologist Kenneth Clark and Gunnar Myrdal (1965), who suggest in *Dark Ghetto: Dilemmas of Social Power* that delinquency is rooted in feelings of alienation and powerlessness among ghetto residents. The solution, according to Clark, was to engage in community organizing to gain power for the poor. Haryou-Act met sharp resistance from city officials, who labeled the staff as corrupt and infiltrated by Communists.

Both MFY and Haryou-Act received substantial operating funds. Mobilization for Youth received approximately $2 million a year, Haryou-Act received about $1 million a year, and 14 similar projects received more than $7 million from the federal Office of Juvenile Delinquency.[16] It was significant that, for the first time, the federal government was pumping large amounts of money into the delinquency prevention effort. Despite intense resistance to these efforts in most cities because local public officials felt threatened, the basic model of Mobilization for Youth was incorporated into the community-action component of the War on Poverty.

In 1967, when social scientists and practitioners developed theories of delinquency prevention for President Lyndon Johnson's Crime Commission, MFY was still central to their thinking (President's Commission on Law Enforcement and the Administration of Justice, 1967). Their problem was to retain a focus upon delivery of remedial services in education, welfare, and job training to the urban poor without creating the intense political conflict engendered by the community action approach.

The issue was complicated because leaders such as Malcolm X and Cesar Chavez, and groups such as the Black Muslims and the Black Panther Party, articulated positions of self-determination and community control. These proponents of ethnic pride and "power to the people" argued that welfare efforts controlled from outside were subtle forms of domestic colonialism. The riots of the mid-1960s dramatized the growing gap between people of color in the United States and their more affluent "benefactors."

It is against this backdrop of urban violence, a growing distrust of outsiders, and increased community-generated self-help efforts that delinquency prevention efforts of the late 1960s and early 1970s developed. A number of projects during this period attempted to reach the urban poor who had been actively involved in ghetto riots during the 1960s. In Philadelphia, members of a teenage gang were given funds to make a film and start their own businesses. Chicago youth gangs such as Black P. Stone Nation and the Vice Lords were subsidized by federal funding, the YMCA, and the Sears Foundation. In New York City, a Puerto Rican youth group, the Young Lords, received funds to engage in self-help activities. In communities across the nation there was a rapid development of summer projects in recreation, employment, and sanitation to help carry an anxious white America through each potentially long, hot summer. Youth patrols were even organized by police departments to employ ghetto youths to "cool out" trouble that might lead to riots. Few of the programs produced the desired results and often resulted in accusations of improperly used funds by the communities. Often, financial audits and investigations were conducted to discredit community organizers and accuse them of encouraging political conflicts with local officials.

One proposed solution that offered more possibility of controlled social action to benefit the young was the Youth Service Bureau (YSB; Norman, 1972). The first YSBs were composed of people from the communities and representatives of public agencies who would hire professionals to deliver a broad range of services to young people. The central idea was to promote cooperation between justice and welfare agencies and the local communities. Agency representatives were expected to contribute partial operating expenses for the programs, and together with neighborhood representatives, decide on program content. Proponents of the YSB approach stressed the need for diverting youthful offenders from the criminal justice system and for delivering necessary social services to deserving children and their families. Ideally, YSBs were designed to increase public awareness of the need for more youth services.

The YSBs generally met with poor results. Intense conflict often arose between community residents and agency personnel over the nature of program goals, and YSBs were criticized for not being attuned to community needs (Duxbury, 1972; U.S. Department of Health, Education, and Welfare, 1973). Funds for these efforts were severely limited in relation to the social problems they sought to rectify. In some jurisdictions, YSBs were controlled by police or probation departments, with no direct community input. These agency-run programs temporarily diverted youths from entering the criminal justice process by focusing on services such as counseling.

The most important aspect of the YSBs was their attempts to operationalize the **diversion** of youth from the juvenile justice process, although the effort's success seems highly questionable. Some argue that diversion programs violate the legal rights of youths, as they imply a guilty plea. Others warn that diversion programs expand the welfare bureaucracy because youths who once would have simply been admonished and sent home by police are now channeled into therapeutic programs. Still others believe that diversion without social services does not prevent delinquency. In any case, a major shift has occurred from the community participation focus of the Mobilization for Youth to a system in which community inputs are limited and carefully controlled. This change in operational philosophy often is justified by the need to secure continued funding, as well as by claims of increasing violence by delinquents. It is important to remember, however, that these same rationales were used to justify a move away from the community organizing model of the Chicago Area Project of the 1930s. Whenever residents become involved in decision making, there are inevitably increased demands for control of social institutions affecting the community. Such demands for local autonomy question the existing distributions of money and power and thus challenge the authority of social control agencies.

INSTITUTIONAL CHANGE AND COMMUNITY-BASED CORRECTIONS

Correctional institutions for juvenile delinquents were subject to many of the same social and structural pressures as community prevention efforts. For instance, there was a disproportionate increase in the number of youths in correctional facilities as blacks migrated to the North and the West. In addition, criticism of the use of juvenile inmate labor, especially by organized labor, disrupted institutional routines. But throughout the late 1930s and the 1940s, increasing numbers of youths were committed to institutions. Later on, the emergence of ethnic pride and calls for black and brown power would cause dissension within the institutions.

The creation of the California Youth Authority (CYA) just prior to World War II centralized the previously disjointed California correctional institutions.[17] During the 1940s and 1950s, California, Wisconsin, and Minnesota developed separate versions of the Youth Authority concept. Under the Youth Authority model, criminal courts committed youthful offenders from 16 to 21 years old to an administrative authority that determined the proper correctional disposition.[18] The CYA was responsible for all juvenile correctional facilities, including the determination of placements and parole. Rather than reducing the powers of the juvenile court judge, the CYA streamlined the dispositional process to add administrative flexibility. The CYA was introduced into California at a time when detention facilities were overcrowded, institutional commitment rates were rising, and the correctional system was fragmented and compartmentalized (Macallair, 2015).

The CYA model was developed by the American Law Institute, which drew up model legislation and lobbied for its adoption in state legislatures. The American Law Institute is a nonprofit organization that seeks to influence the development of law and criminal justice. The institution is oriented toward efficiency, rationality, and effectiveness in legal administration.

The treatment philosophy of the first youth authorities was similar to the approach of William Healy and the child guidance clinic. John Ellingston, formerly chief legislative lobbyist for the American Law Institute in California, related a debate between Healy and Clifford Shaw over the theoretical direction the new Youth Authority should follow. The legislators, persuaded by Healy's focus on diagnosis of individual delinquents, ensured that the clinic model became the dominant approach in California institutions.

Sociologist Edwin Lemert attributed the emergence of the CYA to the growth of an "administrative state" in the United States. In support of this assertion, Lemert noted the trend toward more centralized delivery of welfare services and increased government regulation of the economy, together with the "militarization" of U.S. society produced by war. Lemert, however, did not discuss whether the purpose of this administrative state was to preserve the existing structure of privilege. The first stated purpose of the CYA was "to protect society by substituting training and treatment for retributive punishment of young persons found guilty of public offenses" (Lemert & Rosenberg, 1948, pp. 49–50).

The centralization of youth correction agencies enabled them to claim the scarce state delinquency prevention funds. In-house research units publicized the latest treatment approaches. In the 1950s and the 1960s, psychologically oriented treatment approaches, including guided group interaction and group therapy, were introduced in juvenile institutions. During this period of optimism and discovery, many new diagnostic and treatment approaches were evaluated. Correctional administrators and social scientists hoped for a significant breakthrough in treatment, but it never came. Although some questionable evaluation studies claimed successes, there is no evidence that the new therapies had a major impact on recidivism. In fact, some people began to question the concept of enforced therapy and argued that treatment-oriented prisons might be more oppressive than more traditional institutional routines (Mathieson, 1965). Intense objections have been raised particularly against drug therapies and behavior modification programs. Takagi (1974) views this as the period when brainwashing techniques were first used on juvenile and adult offenders.[19]

Another major innovation of the 1960s was the introduction of community-based correctional facilities. The central idea was that rehabilitation could be accomplished more effectively outside conventional correctional facilities. This led to a series of treatment measures such as group homes, partial release programs, halfway houses, and attempts to decrease commitment rates to juvenile institutions. California was particularly active in developing community-based correctional programming. The Community Treatment Project, designed by Marguerite Warren in California, was an attempt to replace institutional treatment with intensive parole supervision and psychologically oriented therapy. Probation subsidy involved a bold campaign by CYA staff to convince the state legislature to give cash subsidies to local counties to encourage them to treat juvenile offenders in local programs. Probation subsidy programs were especially oriented toward strengthening the capacity of county probation departments to supervise youthful offenders.[20]

Proponents of the various community-based programs argued that correctional costs could be reduced and rehabilitation results improved in a community context. Reducing state expenditures became more attractive as state governments experienced the fiscal crunch of the late 1960s and the 1970s.[21] It also was thought that reducing institutional populations would alleviate tension and violence within the institutions, but it appears that these community alternatives have created a situation in which youngsters who are sent to institutions are perceived as more dangerous, and as a result, are kept in custody for longer periods of time.

The ultimate logic of the community-based corrections model was followed by the Department of Youth Services in Massachusetts, which closed all of its **training schools** for delinquents. Youngsters were transferred to group home facilities, and services were offered to individual children on a community basis (Bakal, 1973). The Massachusetts strategy met intense public criticism by juvenile court judges, correctional administrators, and police officials. Some recent attempts have been made to discredit this policy and to justify continued operation of correction facilities, but the Massachusetts strategy has influenced a move to deinstitutionalize children convicted of status offenses—offenses that are considered crimes only if committed by children, such as truancy, running away, or incorrigibility. In 1975, the federal government made $15 million available to local governments that developed plans to deinstitutionalize juvenile status offenders.

At the moment, the forces opposing institutionalized care are making ideological headway due to past failures of institutional methods in controlling delinquency. Many of these positive changes are the direct result of litigation challenging the treatment of youth in correctional facilities. The costs of maintaining constitutionally acceptable facilities have risen dramatically. It remains uncertain if the powerful have actually embraced a new philosophic approach to juvenile crime or if the current reforms are mostly reactive (Krisberg, 2016). However, previous experience suggests that the pendulum is likely to swing back in favor of the institutional approaches if intense fear of juvenile crime reemerges. Already there is increased talk about the "senseless violence" by youths and the alleged increase in the number of teenage killers; these words have always signaled the beginning of an ideological campaign to promote more stringent control measures and extended incarceration or detention. It is very significant that many states are closing the worst training schools, but the elaboration of well-funded and enriched community-based treatment is still in the beginning stages (Krisberg, 2016). Most jurisdictions still rely on placement in institutions for serious and chronic offenders, with conditions reminiscent of the reform schools of 100 years ago. Many children continue to be warehoused in large correctional facilities, receiving little care or attention. Eventually, they are returned to substandard social conditions to survive as best they can.

CHANGES IN JUVENILE COURT LAW

In the late 1960s, the growing awareness of the limitations of the juvenile justice system resulted in a series of court decisions that altered the character of the juvenile court. In *Kent v. United States* (1966), the Supreme Court warned juvenile courts against "procedural arbitrariness," and in *In re Gault* (1967), the court recognized the rights of juveniles in such matters as notification of charges, protection against self-incrimination, the right to confront witnesses, and the right to have a written transcript of the proceedings. Justice Abe Fortas wrote, "Under our Constitution the condition of being a boy does not justify a kangaroo court" (*In re Gault*, 1967). The newly established rights of juveniles were not welcomed by most juvenile court personnel, who claimed that the informal humanitarian court process would be replaced by a junior criminal court.

Communities struggled with methods of providing legal counsel to indigent youth and with restructuring court procedures to conform to constitutional requirements.

The principles set forth in *Kent,* and later in the *Gault* decision, offer only limited procedural safeguards to delinquent youth (Kittrie, 1971). Many judicial officers believe the remedy to juvenile court problems is not more formality in proceedings, but more treatment resources. In *McKiever v. Pennsylvania* (1971), the Supreme Court denied that jury trials were a constitutional requirement for the juvenile court. Many legal scholars believe the current Supreme Court has a solid majority opposing extension of procedural rights to alleged delinquents. The dominant view is close to the opinion expressed by Chief Justice Warren Burger in the *Winship* case:

> What the juvenile court systems need is less not more of the trappings of legal procedure and judicial formalism; the juvenile court system requires breathing room and flexibility in order to survive the repeated assaults on this court. The real problem was not the deprivation of constitutional rights but inadequate juvenile court staffs and facilities. (*In re Winship,* 1970)

The Supreme Court's decision in *Schall v. Martin* (1984) signaled a much more conservative judicial response to children's rights. Plaintiffs in *Schall v. Martin* challenged the constitutionality of New York's Family Court Act as it pertained to the preventive detention of juveniles. It was alleged that the law was too vague and that juveniles were denied due process. A federal district court struck down the statute and its decision was affirmed by the U.S. Court of Appeals. However, the U.S. Supreme Court reversed the lower courts, holding that the preventive detention of juveniles to protect against future crimes was a legitimate state action.

In the first decades of the 21st century, the judicial view of young people who commit serious crimes is undergoing tremendous change. In 2005, the U.S. Supreme Court banned the use of **capital punishment** for persons who had committed their crimes before the age of 18 (*Roper v. Simmons*). The decision was based on a 5–4 vote. Writing for the majority, Justice Kennedy cited international standards that banned the death penalty for juveniles. He also argued that there was an emerging consensus against the practice, and he cited new research on **neuroscience** that demonstrated that persons under age 18 were generally incapable of understanding the future consequences of their behavior. While there had been a previous Supreme Court ban on executing youth under age 17 for states that had no minimum **age of jurisdiction** (*Thompson v. Oklahoma,* 1987), the *Roper* case went further.

The logic of the Supreme Court's decision in *Roper* was extended in 2010 to prohibit persons under age 18 from being sentenced to life without the possibility of parole (LWOP) if their crimes were not homicides. But in 2012, the U.S. Supreme Court declared that LWOP sentences violated the **Eighth Amendment** even for youth convicted of homicide (*Jackson v. Hobbs,* 2012*)*. The march toward a new jurisprudence for youthful offenders continued in *Miller v. Alabama* (2012) in which convicts who had received LWOP or "virtual life" sentences could petition for a habeas corpus hearing to determine if the original sentencing court had appropriately considered their level of immaturity and emotional vulnerability at the time of the crime.

The increased legal understanding of the special needs of young people also was expressed in a U.S. Supreme Court case in *J.D.B v. North Carolina* (2011), where the court determined that age must be taken into consideration when defining the status of "custody" that requires law enforcement to administer a Miranda warning. In this case, a 13-year-old special education student was questioned in a conference room by a uniformed officer and two school officials. He was not informed of his rights, and his guardians were not notified. The young person confessed to burglaries. Only after the self-incrimination occurred was the child informed that he could leave.

Justice Sotomayor opined that children are "generally less mature and responsible than adults." The justice noted that young people "often lack the experience, perspective and judgment to recognize and avoid choices that could be detrimental to them" (*J.B.D. v. North Carolina*, 2011). The court's majority opinion concluded that the individual's age must be a factor when deciding if the youngster was "in custody."

There is also growing political and judicial consensus that holding young people in solitary confinement may also violate the Eighth Amendment prohibition on **cruel and unusual** punishment. President Barack Obama has instructed the Federal Bureau of Prisons to end this practice (Obama, 2016). A prestigious taskforce led by the U.S. Deputy Attorney General has forcefully urged the ending of solitary confinement of young people in federal and state corrections facilities (U.S. Department of Justice, 2016). Several states have also effectively ended solitary confinement for juveniles (Daugherty, 2015).

In addition to Supreme Court decisions, the U.S. Congress has enacted several laws that aim to reduce the abuse of juveniles in correctional facilities. Advocates for incarcerated young people are increasing these statutory remedies to challenge harsh conditions of confinement in youth institutions. In 1980, Congress enacted and the president signed into law the Civil Rights of Institutionalized Persons Act (CRIPA). This law gave powers to the U.S. Justice Department Civil Rights Division to investigate and bring actions against state and local government that were violating the rights of adults and juveniles in a wide range of institutional settings. These actions seek reforms of systemic abuses as opposed to the claims of individual inmates. Since its passage, there have been hundreds of facilities that have been investigated. Juvenile facilities have been subject to these reviews in California, Indiana, Illinois, New York, Arizona, Oklahoma, Mississippi, and the District of Columbia. Topics included in these investigations include excessive violence in these institutions, abusive practices by staff, lack of treatment services, especially for mentally ill youth, extensive uses of solitary confinement as punishment, and the denial of education, health care, and counseling services. Most of these CRIPA investigations result in agreements to implement the requested reforms.

The Prison Rape Elimination Act (PREA) of 2003 was designed to protect vulnerable persons from physical and sexual exploitation in adult and juvenile facilities. It also banned traditional practices that discriminated against LGBT persons, especially policies of isolating such persons in solitary confinement. National research has suggested that young people are victimized in both juvenile and adult corrections facilities. The PREA requires careful monitoring of facilities by outside auditors as well as confidential methods by which youth and adult inmates can report abusive practices.

Another key legislative action was the Americans with Disabilities Act (ADA) in 1990. Both youth and adults with physical or mental impairments are entitled to protection and oversight by the U.S. Justice Department. This scrutiny is designed to end discrimination against disabled persons and to promote their full integration into mainstream activities. Persons identified as disabled are entitled to reasonable accommodations in "all aspects of their correctional programs." All corrections systems are required to record all inmates who experience significant impairments of important life functions and to attempt to remedy or alleviate those limitations. Young people also have additional rights under the Individuals with Disabilities Education Act (IDEA) for appropriate special education services. The courts are increasingly agreeing that many routine correctional practices, including solitary confinement and standardized disciplinary systems, violate youth rights under the ADA and IDEA.

These large expansions for protected rights for incarcerated young people led to political blowback that claimed that the new inmate rights were excessive and were endangering the safety of staff and the general public. The clearest example of this counterattack was the Prison Reform Litigation Act (PRLA) of 1995. The PRLA was promoted by conservatives as a solution to the alleged wave of frivolous civil rights litigation by prisoners. It limited the reasons why inmates could ask for relief in the federal courts, banning claims for psychological and emotional injuries. It required that convicts prove that they had exhausted all available administrative remedies and capped the amount of legal fees for attorneys who represented indigent inmates. The PRLA also required that judges show that the remedies they ordered were the least drastic options. Further, the courts had to demonstrate that efforts to reduce severe prison crowding would not endanger public safety or the safety of prison workers.

The PRLA had a chilling effect on the number of civil rights actions challenging the maltreatment of inmates. The processes set up to review *compliance* with the PRLA ultimately led to a landmark Supreme Court decision in *Brown v. Plata et al.* (2011) that broadens the court's interpretation of the Eighth Amendment prohibition against cruel and usual punishment. Some of the language of the PRLA has been used to encourage courts to remedy gross deficiencies in juvenile facilities via **consent decrees** (*G.F. et al. v. Contra Costa County*, 2015; *R.J. et al. v. Bishop*, 2009; *Farrell v. Brown*, 2005).

THE EMERGENCE OF A CONSERVATIVE AGENDA FOR JUVENILE JUSTICE

The PRLA and the political reaction to judicial and legislative efforts to improve the treatment of incarcerated young people was part of the ongoing legacy of a conservative agenda that dominated the field of juvenile justice for almost two decades. From the late 1970s and into the 1980s, a conservative reform agenda dominated the national debates over juvenile justice. This new perspective emphasized **deterrence** and punishment as the major goals of the juvenile court. Conservatives called for the vigorous prosecution of serious and violent youthful offenders. They alleged that the juvenile court was overly lenient with dangerous juveniles.

Conservatives also questioned the wisdom of diverting status offenders from secure custody. The Reagan administration introduced new programs in the areas of missing children and child pornography, which were problems allegedly created by the liberal response to status offenders. Substantial amounts of federal funds were spent on police intelligence programs and enhanced prosecution of juvenile offenders.

Changes in federal policy were also reflected in the actions of many state legislatures. Beginning in 1976, more than half the states made it easier to transfer youths to adult courts. Other states stiffened penalties for juvenile offenders via mandatory minimum sentencing guidelines.

The most obvious impact of the conservative reform movement was a significant increase in the number of youths in juvenile correctional facilities. In addition, from 1979 to 1984, the number of juveniles sent to adult prisons rose by 48%. By 1985, the Bureau of Justice Statistics reported that two thirds of the nation's training schools were chronically overcrowded.

Another ominous sign was the growing proportion of minority youth in public correctional facilities. In 1982, more than one half of those in public facilities were minority youths, whereas two thirds of those in private juvenile facilities were white. Between 1979 and 1982, when the number of incarcerated youth grew by 6,178, minority youth accounted for 93% of the increase. The sharp rise in incarceration occurred even though the number of arrests of minority youth declined. As noted in Chapter 1, the renewed push to punish and incarcerate young people was fueled by a "moral panic" about the dangers posed by "crack babies" and the phony claims of a new dangerous class that was labeled as "superpredators" by politicians of varying ideological positions. This "war against children" was tragically reminiscent of the myth-driven political campaigns that led to the houses of refuge and the child-saving movement of the 19th century.

The rise in the number of youth in confinement was accompanied with efforts to send more youngsters into adult prisons and jails. There were also legislative actions to reduce the amount of education and recreation provided to youth in custody. States such as Illinois, California, Texas, Florida, New York, and Michigan transferred their juvenile corrections systems to the management of the adult corrections agency. This change led to an extraordinary decline in the living conditions and treatment of incarcerated youth. As noted above, this collapse of the traditional values of the juvenile justice system led to key litigation and additional federal civil rights actions. The net impact of this new wave of litigation on behalf of youth rights produced a major increase in the cost of incarceration. It is not uncommon for state juvenile facilities to spend over $200,000 to lock up one youth for a year. The costs of defending lawsuits and the tremendous rise in juvenile corrections costs led many states and localities to embark on major efforts to dramatically reduce the numbers of young people growing up behind bars. Between 1999 and 2011, the number of youth in residential placements declined by 43% (Sickmund & Puzzanchera, 2014). The largest decline was observed in California state juvenile facilities that decreased from over 10,000 inmates in the late 1990s to less than 500 today. These young people were not simply transferred to adult facilities. There were also significant declines in the number of young people confined in county facilities. Similar large declines also occurred in almost every state (Krisberg, Bakal, DeMuro, & Schiraldi, 2015; Krisberg, 2016). It is worth remembering that despite this major decline in youth incarceration, the arrest rate of juveniles continued a steady drop across the nation.

The reduction in youth incarceration was not a direct result of improving educational and employment prospects for young people, nor did states and localities increase their investments in vulnerable families and their children. It appears that the dominant fiscal response was to cut public investment in community programs while attempting to preserve public employee jobs, pensions, and salaries. This is an unsustainable formula that was paid for by more taxation hidden as "fees" and cuts in state spending for public education, especially colleges and universities. These practices led to greater voter alienation from "established" politics but also has ratcheted up racism, xenophobia, and resentment against the poor.

SUMMARY

We have traced the history of the juvenile justice system in the United States in relation to significant population migrations, rapid urbanization, race conflicts, and transformation in the economy. These factors continue to influence the treatment of children. The juvenile justice system traditionally has focused on the alleged pathological nature of delinquents, ignoring how the problems of youths relate to larger political and economic issues. Both institutional and community-based efforts to rehabilitate delinquents have been largely unsuccessful. Those with authority for reforming the juvenile justice system have traditionally supported and defended the values and interests of the well to do. Not surprisingly, juvenile justice reforms have inexorably increased state control over the lives of the poor and their children. The central implication of this historical analysis is that the future of delinquency prevention and control will be determined largely by ways in which the social structure evolves.[22] It is possible that this future belongs to those who wish to advance social justice on behalf of young people rather than to accommodate the class interests that have dominated this history (Krisberg, 1975; Liazos, 1974). However, one must be cautious about drawing direct inferences for specific social reforms from this historical summary. William Appleman Williams (1973) reminds us, "History offers no answers per se, and it only offers a way of encouraging people to use their own minds to make their own history" (p. 480).

REVIEW QUESTIONS

1. Describe the "moral panic" that led to the founding of the houses of refuge. Who were the leaders of this movement, and what did they hope to accomplish?

2. What led to the creation of the first juvenile court in America in 1899?

3. How did legal decisions shape the contours of juvenile justice in the 1960s?

4. How did the philosophy of the Chicago Area Project differ from that of the Child Guidance Clinic movement?

5. What forces led to the decline in youth incarceration in the early decades of the 21st century?

NOTES

1. Thorsten Sellin, *Pioneering in Penology,* provides an excellent description of the Amsterdam House of Corrections.

2. This issue is well treated by Winthrop Jordan in *The White Man's Burden.*

3. Sources of primary material are N. R. Yetman, *Voices from Slavery,* and Gerda Lerner, *Black Women in White America.* Another fascinating source of data is Margaret Walker's historical novel *Jubilee.*

4. Historical data on the 19th century rely on the scholarship of Robert Mennel, *Thorns and Thistles*; Anthony Platt, *The Child Savers: The Invention of Delinquency*; Joseph Hawes, *Children in Urban Society: Juvenile Delinquency in Nineteenth Century America*; and the document collection of Robert Bremner et al. in *Children and Youth in America: A Documentary History.*

5. *Delinquent children* are those in violation of criminal codes, statutes, and ordinances. *Dependent children* are those in need of proper and effective parental care or control but having no parent or guardian to provide such care. *Neglected children* are destitute, are unable to secure the basic necessities of life, or have unfit homes due to neglect or cruelty.

6. A good description of anti-Irish feeling during this time is provided by John Higham, *Strangers in the Land.*

7. The preoccupation with the sexuality of female delinquents continues today. See Meda Chesney-Lind, "Juvenile Delinquency: The Sexualization of Female Crime."

8. This routine is reminiscent of the style of 18th-century American Indian schools. It represents an attempt to re-create the ideal of colonial family life, which was being replaced by living patterns accommodated to industrial growth and development.

9. The term *dangerous classes* was coined by Charles Loring Brace in his widely read *The Dangerous Classes of New York* and *Twenty Years Among Them.*

10. The classic of these studies is that of E. C. Wines, *The State of Prisons and Child-Saving Institutions in the Civilized World,* first printed in 1880.

11. Platt, *The Child Savers: The Invention of Delinquency,* pp. 101–136, and Hawes, *Children in Urban Society: Juvenile Delinquency in Nineteenth Century America,* pp. 158–190, provide the most thorough discussions of the origins of the first juvenile court law.

12. Roy Lubove, *The Professional Altruist,* is a good discussion of the rise of social work as a career.

13. A few earlier clinics specialized in care of juveniles, but these mostly dealt with feeble-minded youngsters.

14. Anthropometric measurements assess human body measurements on a comparative basis. A popular theory of the day was that criminals have distinctive physical traits that can be scientifically measured.

15. Longitudinal studies analyze a group of subjects over time.

16. By comparison, the Chicago Area Project operated on about $283,000 a year.

17. John Ellingston, *Protecting Our Children from Criminal Careers,* provides an extensive discussion of the development of the California Youth Authority.

18. California originally set the maximum jurisdictional age at 23 years but later reduced it to 21. Some states used an age limit of 18 years so that they dealt strictly with juveniles. In California, both juveniles and adults were included in the Youth Authority model.

19. Paul Takagi, in "The Correctional System," cites Edgar Schein, "Man Against Man: Brainwashing," and James McConnell, "Criminals Can Be Brainwashed—Now," for candid discussions of this direction in correctional policy.

20. Paul Lerman, *Community Treatment and Social Control,* is a provocative evaluation of the Community Treatment Project and Probation Subsidy.

21. See James O'Connor, *The Fiscal Crisis of the State,* for a discussion of the causes of this fiscal crunch.

22. This perspective is similar to that of Rusche and Kirchheimer in their criminological classic *Punishment and Social Structure.*

INTERNET RESOURCES

The National Criminal Justice Reference Service is a comprehensive collection of research and materials on the juvenile justice system.
http://www.ncjrs.gov

The American Bar Association has excellent material on the history and evolution of juvenile justice.
http://www.americanbar.org

The Public Broadcasting Service has produced several excellent documentaries about the history of juvenile justice in America.
http://www.pbs.org

Chapter 4

THE CURRENT JUVENILE JUSTICE SYSTEM

The goal of this chapter is to describe how the historical legacy of juvenile justice has been translated into the modern justice system practices that one encounters today. The approach is to describe the processing of cases through the modern juvenile court system, comparing and contrasting these practices with the adult criminal justice system. Keep in mind that there remains significant variation among states and localities. What is presented below is an attempt to capture the present reality in the largest number of juvenile court jurisdictions. The National Center for Juvenile Justice is the preeminent source of more current information on how legislative changes impact the structure and process of the juvenile justice system (Snyder & Sickmund, 1999).

The original scope or jurisdiction of the juvenile justice system is generally defined in each state by specific statutes. The first level of this definition process involves the age of the youth at the time of arrest or referral to the juvenile justice system. Most states set the threshold upper age of juvenile court jurisdiction at 17. The importance of this upper age is to define the presumptive age at which young people are tried in criminal courts as opposed to family or juvenile courts. New York and North Carolina have a much lower upper limit for adolescents in criminal courts at 15 years of age. Twelve other states employ age 16 as the upper limit of original juvenile court jurisdiction. However, several states permit the juvenile court to retain its jurisdiction of cases already in the system until age 21. Several states actually allow this continuing jurisdiction for some youths as old as 25 who are under the control of the state youth corrections authority (Sickmund & Puzzanchera, 2014). Currently, there is a serious reexamination of the traditional age boundaries of the juvenile court. These policy debates are influenced by emerging scientific evidence about the brain development of adolescents. Since 1975, five states have altered the juvenile court age boundaries. Alabama raised the upper age limit from 15 to 16, and then to 17 the next year. Wyoming lowered its upper age limit from 18 to 17, New Hampshire and Wisconsin lowered their limits to 16, and Connecticut enacted legislation that gradually raised the age of jurisdiction from 15 to 17 by 2012.

There is also variability among states in the lower boundary of the juvenile court's authority—the age below which the child is deemed too young to be subject to juvenile court laws. There are sixteen states that define the youngest age at which children can be tried in juvenile courts. Eleven states set this lower age boundary at 10 years old, whereas another five states use boundaries of age 6 (North Carolina), age 7 (Maryland, Massachusetts, and New York), and age 8 (Arizona). Other jurisdictions have no statutorily mandated lower age threshold, although the usual practice is to recognize the tradition of the English Common Law standard of age 7 for the lowest age of criminal court jurisdiction (Snyder & Sickmund, 1999).

These general age limits are further refined by state laws specifying circumstances and legal mechanisms through which the original jurisdiction of the juvenile court may revert to the criminal court or be shared with the criminal court system. The overall trend in juvenile law in the past several decades has been to extend these exceptions and to extend the range of transfer procedures; however, most recently, a few states have sought to restrict the transfer of young offenders to criminal courts or make the standards for the transfers more stringent. For example, Illinois enacted tough laws that permitted prosecutors to automatically send very young adolescents to criminal processing, but a recent law proposed by the governor requires that all young people who were arrested before age 16 are entitled to a juvenile court hearing to determine the appropriateness of their transfer to adult criminal courts. Some would say that the most important development in contemporary juvenile court law has been the "**blurring of the line**" between the juvenile court and the criminal law system as well as newer efforts to redraw these historic lines. There are also suggestions to consider hybrid systems of "**blended sentencing**" in which a youngster can be simultaneously under the authority of the juvenile court and criminal laws of the state.

A further complication in understanding the juvenile justice system is the overlap, in many locales, between the criminal law system as applied to minors and the laws governing the protection of dependent, neglected, and abused children. It is common that children appearing before the juvenile court are still under the auspices of child welfare and protection laws. Judges sometimes use either legal structure or a combination of the two systems to fashion remedies in individual cases. Often, it is the same judge who hears cases in both systems. Moreover, many states have established special legal categories such as Children in Need of Supervision, Persons in Need of Supervision, or Minors in Need of Supervision that overlap the traditional categories of **dependent children** with behavior that might be considered criminal law violations in other states. Most often, these mixed categories are meant to be used in cases involving behaviors such as truancy, running away, curfew violations, underage drinking or smoking, or incorrigibility. Traditionally, many state laws prohibit the secure confinement of these nonoffenders or forbid their confinement with youngsters charged with criminal law violations. However, these finite legal categories often merge together in cases in which the youth is already under juvenile court control (e.g., under probation supervision or in some sort of diversion program). In the world of contemporary juvenile justice, nothing is simple and consistent.

Before discussing the various components of the juvenile justice process, it is worth considering how the modern juvenile and criminal justice systems differ and converge. Table 4.1, prepared by the federal Office of Juvenile Justice and Delinquency Protection, provides a convenient summary of these common and differing aspects of the two justice systems (Snyder & Sickmund, 1999).

At the most basic level, the two systems differ in their core assumptions. For example, the juvenile justice system assumes that young people are capable of positive change. Rehabilitation

Table 4.1   Comparison of Juvenile and Criminal Justice Systems

Although the juvenile and criminal justice systems are more alike in some jurisdictions than in others, generalizations can be made about the distinctions

Juvenile Justice System	Common Ground	Criminal Justice System
Operating Assumptions		
• Youth behavior is malleable. • Rehabilitation is usually a viable goal. • Youths are in families and not independent.	• Community protection is a primary goal. Law violators must be held accountable. • Constitutional rights apply.	• Sanctions should be proportional to the offense. • General deterrence works. Rehabilitation is not a primary goal.
Prevention		
• Many specific delinquency prevention activities (e.g., school, church, recreation) are used. • Prevention is intended to change individual behavior and is often focused on reducing risk factors and increasing protective factors in the individual, family, and community.	• Educational approaches are taken to specific behaviors (drunk driving, drug use).	• Prevention activities are generalized and are aimed at deterrence (e.g., Crime Watch).
Law Enforcement		
• Specialized "juvenile" units are used. • Some additional behaviors are prohibited (truancy, running away, curfew violations). • Some limitations are placed on public access to information. • A significant number of youth are diverted away from the juvenile justice system, often into alternative programs.	• Jurisdiction involves the full range of criminal behavior. • Constitutional and procedural safeguards exist. • Both reactive and proactive approaches (targeted at offense types, neighborhoods, etc.) are used. • Community policing strategies are employed.	• Open public access to all information is required. • Law enforcement exercises discretion to divert offenders out of the criminal justice system.
Intake—Prosecution		
• In many instances, juvenile court intake, not the prosecutor, decides what cases to file. • The decision to file a petition for court action is based on both social and legal factors. • A significant portion of cases are diverted from formal case processing. • Intake or the prosecutor diverts cases from formal processing to services operated by the juvenile court, prosecutor's office, or outside agencies.	• Probable cause must be established. • The prosecutor acts on behalf of the state.	• Plea bargaining is common. • The prosecution decision is based largely on legal facts. • Prosecution is valuable in building history for subsequent offenses. • Prosecution exercises discretion to withhold charges or divert offenders out of the criminal justice system.

(Continued)

Table 4.1   (Continued)

Juvenile Justice System	Common Ground	Criminal Justice System
Detention—Jail/Lockup		
• Juveniles may be detained for their own protection or the community's protection. • Juveniles may not be confined with adults unless there is sight-and-sound separation.	• Accused offenders may be held in custody to ensure their appearance in court. • Detention alternatives of home or electronic detention are used.	• Accused individuals have the right to apply for bond/bail release.
Adjudication—Conviction		
• Juvenile court proceedings are quasi-civil (not criminal) and may be confidential. • If guilt is established, the youth is adjudicated delinquent regardless of offense. • Right to jury trial is not afforded in all states.	• Standard of proof beyond a reasonable doubt is required. • Rights to be represented by an attorney, to confront witnesses, and to remain silent are afforded. • Appeals to a higher court are allowed. • Experimentation with specialized court (i.e., drug courts, gun courts, and reentry courts).	• Defendants have a constitutional right to a jury trial. • Guilt must be established on individual offenses charged for conviction. • All proceedings are open.
Disposition—Sentencing		
• Disposition decisions are based on individual and social factors, offense severity, and youth's offense history. • Dispositional philosophy includes a significant rehabilitation component. • Many dispositional alternatives are operated by the juvenile court. • Dispositions cover a wide range of community-based and residential services. • Disposition orders may be directed to people other than the offender (e.g., parents). • Disposition may be indeterminate, based on progress demonstrated by the youth.	• Decisions are influenced by current offense, offending history, and social factors. • Decisions hold offenders accountable. Decisions may give consideration to victims (e.g., restitution and no-contact orders). • Decisions may not be cruel or unusual.	• Sentencing decisions are bound primarily by the severity of the current offense and by the offender's criminal history. • Sentencing philosophy is based largely on proportionality and punishment. • Sentence is often determinate, based on offense.
Aftercare—Parole		
• Function combines surveillance and reintegration activities (e.g., family, school, work).	• The behavior of individuals released from correctional settings is monitored. • Violation of conditions can result in incarceration of the youth.	• Function is primarily surveillance and reporting to monitor illicit behavior.

SOURCE: U.S. Department of Justice, Office of Juvenile Justice and Delinquency Prevention.

is considered a possible and desirable goal. Juvenile justice interventions assume that youths are part of families, and attempts are made to involve parents in the various programs of the system. By contrast, the current criminal justice system does not place much emphasis on rehabilitation as a goal, although many criminal law systems did value rehabilitation in the past. The contemporary criminal justice system assumes that general deterrence is an effective crime control strategy (Zimring & Hawkins, 1973). This means that the criminal law assumes that others will be discouraged from law breaking by knowledge of the severity and certainty of punishments meted out to apprehended offenders. Inherent in the concept of deterrence is the assumption that humans operate principally through a process of rational calculations of the benefits of committing crimes compared to the costs of apprehension and punishment (Wilson, 1983). The juvenile justice system assumes that in addition to cost–benefit calculations, young offenders are motivated by a range of rational and irrational forces. The reader will recall that this latter assumption is directly related to the early scientific work supporting the creation of the juvenile court (Healy, 1915).

Both the juvenile and the criminal justice systems recognize protection of the public as a major goal. In each system there is value placed on holding offenders accountable for their specific actions. Most important, since the late 1960s, the juvenile justice system has mirrored the adult court system in its attention to the protection of the constitutional rights of the accused. Both systems are increasingly interested in advancing the rights of victims in various aspects of its legal processes. Legislation establishing the rights of victims in the juvenile court has trailed behind the criminal court system, but the 1990s was a period in which many lawmakers sought to extend rights to victims within the juvenile justice system.

The two systems also differ markedly in their approaches to crime prevention. As noted earlier, the criminal justice system depends on general and specific deterrence to prevent future crimes. Prevention strategies often entail increasing the capacity and willingness of citizens to report crimes to the police. This type of public participation is reflected in programs such as Crime Watch or Crime Stoppers. Crime Watch programs often involve organizing neighborhood residents to conduct informal patrols, pass along information about recent local crimes, or hold meetings with police to learn how to better protect one's person or property. Crime Stoppers is a national program that pays cash rewards for information leading to the arrest of criminals. Juvenile crime prevention programs tend to rely more on educational efforts (an example is the Drug Abuse Resistance Education program, in which police officers enter schools and teach children about the dangers of drug use) and the delivery of services including recreational, vocational, and counseling efforts. The Office of Juvenile Justice and Delinquency Prevention funds prevention programs through a number of grants to states and localities. Units of state and local governments typically devote some funding each year to service-oriented prevention efforts. These services are designed to reduce certain risk factors that researchers have linked to the onset of delinquent behavior in individuals, families, schools, and communities (Hawkins & Catalano, 1992). Prevention programs also seek to increase protective or buffering factors that insulate youngsters from risk factors. We discuss these approaches to delinquency prevention in Chapter 7. There are some more limited educational prevention efforts in the criminal justice system such as classes for drunk drivers, educational programs for domestic violence offenders, or courses on traffic safety. Participation in these adult prevention programs is often a condition of diverting the case from formal court processing or of reducing penalties.

Law enforcement policies and practices differ between the two justice systems. Police agencies often assign special units to handle juvenile offenders, and there is an expectation that public access to information about juvenile offenders will be more limited than for adults. The overall philosophy of enforcing laws with minors is to divert a significant number of these cases away from the justice system. In policy, and often in practice, the ideal is to handle cases informally, allowing parents or other community agencies to take the lead. The ethos of juvenile justice is to exercise extreme constraint when intervening in the lives of children. Chapter 3 explains why and how these ideas became part of the modern juvenile justice system, and it is true that this non-interventionist paradigm is regularly questioned and reexamined.

It is also worth noting that police respond to behaviors by juveniles that would not engage their attention if committed by adults. For example, police enforce truancy laws but do not arrest adults who fail to attend their classes. Except in times of emergency, adults do not have to adhere to curfews, whereas minors may be restricted in the hours when they can be in public places. Youth curfews may require that the child be accompanied by responsible caretakers. Adults are free to purchase and use highly toxic and addictive substances such as tobacco and alcohol, but these commodities are forbidden for juveniles. Minors can be incarcerated for running away from their homes, whereas adults are free to change their living situation without criminal penalty.

Police divert adult offenders from further criminal justice processing, although the rationale is rarely that there are better places from which to get services. Diversion of adult offenders is often based on police obtaining information from them that can be used to arrest others. The exercise of police discretion in adult cases has led to serious criticism of law enforcement agencies who are alleged to employ racial, gender, or social class factors in making their decisions. For example, police handling of domestic violence incidents has moved toward mandatory arrest policies in lieu of allowing the street officers to decide the appropriate resolution of individual cases. While these mandatory arrest policies have been promoted based on research findings (Sherman et al., 1997), the evidence supporting the efficacy of these approaches is less than totally convincing.

More recently, police practices for adults and juveniles have converged in the theory of problem-solving law enforcement in which informal dispute resolution and mediation strategies are encouraged in lieu of making arrests. For both juveniles and adults, police are governed by common constitutional requirements pertaining to the limitations on powers to stop individuals, rules pertaining to search and seizure, and safeguards against self-incrimination. However, the law has been evolving on how strictly law enforcement must adhere to constitutional protections. As noted earlier, the courts in cases such as *J.B.D v. North Carolina* (2011) are expanding the legal protections for detained youth. The legal rights of juveniles who are detained and questioned in school settings is still a subject of intense debate. One major issue remains that the violations of rights for adult defendants may lead to the exclusion of crucial evidence; for juveniles, they may be processed by the juvenile court and confined for "their own protection" without definite evidence of criminal wrongdoing.

As one moves further along in the process, the juvenile and adult systems begin to diverge. In the adult system, it is the prosecutor who decides if a person who is arrested by the police is brought into the criminal justice system. Typically, there is a preliminary hearing in which the prosecutor presents the charges to a judge, the defendant enters a plea of guilty or not guilty, and either bail is set or the defendant is remanded to jail. Later in the process, usually after a variety of evidentiary motions are considered (such as the inclusion of evidence, the face validity of the charges, or other procedural matters), the prosecutor decides either to bring the case forward or to drop it. In many jurisdictions, a panel of citizens known as a grand jury will hear the prosecutor's preliminary evidence and will decide if the criminal case should be brought forward. Prosecutors have the discretion to either drop the case or defer prosecution, usually in situations in which the defendant agrees to enter some treatment or counseling program. Failure to complete this agreement often results in the original charges being filed.

The juvenile justice system usually employs an intake unit to evaluate whether a case should be brought forward. Although it is increasingly common for prosecutors to participate in the intake process, it is most often accomplished by staff of the juvenile probation department. The sufficiency of legal evidence is a strong consideration, but the primary emphasis of the juvenile intake process is problem solving—how best to rectify the youth's misconduct with a minimum of court intervention. The juvenile court almost always is governed by the principle of utilizing the least restrictive alternative consistent with protecting public safety. The intake process is also guided by a search for ways to protect the child from adverse community, family, or peer influences. It is sometimes the case that the child welfare or mental health systems may provide more appropriate responses to the child's behavior than the juvenile justice system.

In the juvenile court, the decision is made whether to file a petition to declare the child a ward of the court. This decision is guided by both legal and social service considerations. The child and his or her attorney may contest the petition, but they do not typically enter a plea of guilt or innocence. In most locales, a juvenile court judge will decide whether it is in society's or the child's best interests to proceed with a formal hearing on the petition. In smaller jurisdictions, this hearing is conducted by a lower court judge or magistrate. Unlike the criminal justice system, there is no automatic right to bail for juveniles. Youths who are not released to their guardians are confined in juvenile detention facilities. The juvenile court is permitted broad latitude to detain youngsters for society's protection (similar to the adult system) or for the protection of the child. Whereas there are very strict constitutional limitations on preventive pretrial detention for adults, the U.S. Supreme Court has approved a much wider set of detention criteria for minors (*Schall v. Martin*, *1984*).

It is at the adjudication stage that the two justice systems are most different. Juvenile court hearings are not viewed as criminal law proceedings; they are quasi-civil hearings. If the facts of an alleged petition are sustained by the court, the petition is sustained and the youth is declared to be a ward of the juvenile court. In the criminal law system, the offender is found guilty of a very specific set of criminal offenses; in the juvenile court, the petition relates to a general pattern of behavior or family circumstances. Criminal court hearings must be open to the public, whereas juvenile courts operate primarily on a confidential basis. Adults charged with criminal offenses are entitled to a jury of their peers, but there is no legally recognized right to jury trials in the juvenile court system. There has been a trend toward opening up juvenile court hearings to the public. As of 2010, there were 18 states permitting public attendance in hearings, with some rare exceptions for protection of the child. There are another 20 states that permit public access for a limited set of cases. For example, the proceedings might be open in felony cases involving older youth and those with prior court involvement (Sickmund & Puzzanchera, 2014). There also is variation among the states on whether juveniles can be fingerprinted, whether criminal records are sealed from public view, or if schools must be notified of a student's involvement with law enforcement or the courts. In the 1990s, many states moved to reduce the privacy and confidentiality of juvenile court records and proceedings. In the past several years, there have been discussions to roll back some of these "get tough laws" and to give court-processed young people a second chance.

Both the criminal and juvenile justice systems require a standard of proof beyond a reasonable doubt, and persons in each court have rights to be represented by attorneys, to confront their accusers, and to not testify against themselves. The right to appeal is available in both systems.

Both court systems are beginning to introduce specialized sections of their courts that focus on particular offenses such as drug courts, gun courts, and **reentry** courts.

During the dispositional phase of juvenile court hearings, there is great weight placed on in-depth social, psychological, and individual factors. Dispositions are, by design, to be *individualized*. Juvenile courts may even direct their dispositional orders to other individuals such as the youth's guardians. Juvenile courts employ a broad range of dispositional options—orders to make **restitution** or to perform community service, as well as a wide range of residential placements. Juvenile courts often impose curfews or require regular attendance in school. Parents and youths may be ordered to attend counseling sessions. There are more creative dispositions sometimes utilized in criminal courts but on a far more limited basis.

Sentencing in criminal courts generally relies more on a limited set of legal factors such as the gravity of the current offense and the prior record of the offender. Criminal court sentences are very often rigidly determined by state laws, whereas juvenile court dispositions are far more open ended—the youth is declared under the juvenile court's supervision based on regular reviews of his or her progress up to the statutory age maximum jurisdiction of the court. Neither the criminal nor juvenile court system may impose sanctions that are deemed cruel or unusual.

Virtually every state has laws that define the circumstances in which youth may be processed in criminal courts and eligible for adult court sanctions. There are various legal means by which this can happen. Most states define a court-supervised process that permits juvenile court judges to waive their jurisdiction and transfer a child to the criminal court. This process is often referred to as a "certification," "fitness," or "remand." Some states even permit a "reverse **waiver**," allowing the criminal courts to return a defendant to the juvenile court. Other states have specific laws that define the ages and offenses that create the presumption that a young person will be tried in a criminal or juvenile judicial venue. These state laws are frequently changed as elected officials express their sense of public outrage over well-publicized crime committed by young people. Other jurisdictions permit prosecutors to "direct file" youthful defendants to criminal courts without the requirement for judicial review. Many states require that young people who have ever been tried as adults must always be sent to criminal courts. There are also hybrid systems in which either adult or juvenile courts may sentence young people as both adults and juveniles. This practice is known as *blended sentencing*. These hybrid systems offer the benefits of the child protection aspects of the juvenile court laws but can impose harsher criminal penalties if the young person commits a new offense or violates the requirements of the less severe sanction. At present, there is no consistent data or easily available statistics across states on the utilization of these widely varying approaches to juvenile lawbreaking. For example, it has been observed that young people of color are far more likely to be more harshly punished in both the juvenile and adult justice systems, but it is currently impossible to adjudicate this claim at an aggregate level of states or local jurisdictions.

The vast majority of transfers are the result of mandatory statutes and not judicial waiver hearings. Laws governing the upper age of the juvenile court may be responsible for as many as 137,000 youth processed in criminal courts (Sickmund & Puzzanchera, 2014).

The hysteria over the superpredator myth led to a significant expansion of the mandatory laws and other methods of trying more youngsters as if they were adults. These new approaches rarely considered the mental and emotional condition of the youth but were usually determined by the chronological age of the child and the perceived gravity of the alleged offense. As noted earlier, the jurisprudence about age of maturity and criminal responsibility is being actively reviewed and reconsidered by many state legislatures and courts (Zimring, 2005).

The borderline between the adult and juvenile justice systems is even more complex. Most offenses are resolved via **plea bargains** or discretionary decisions made by police, prosecutors, or probation staff. These deliberations are not made in public settings. In fact, we do not really know how many young people are punished via the criminal justice system without a fully transparent legal process.

The Flow of Cases Through the Juvenile Justice System

Figure 4.1: The Stages of Delinquency Case Processing in the Juvenile Justice System (Snyder and Sickmund, 1999, p. 98) shows a typical flow of cases through the juvenile justice system.

It cannot be repeated often enough that there are many state variations that alter how this process operates across the nation. Let's examine this case flow using data from the National Center for Juvenile Justice (Snyder & Sickmund, 1999).

Law enforcement agencies are the primary source of referrals into the juvenile justice system. Approximately 86% of cases come from police agencies; the remaining 14% are referred by schools, social service agencies, or parents. In the case of minor assaults, there are a larger number of referrals coming from schools, probably a reflection of campus fights among students.

Police agencies have a range of policies on handling juvenile arrests. Some departments operate their own special programs that are designed to divert youngsters from further court processing. There are police agencies that work diligently to return back to their parents those juveniles that are stopped for minor offenses. Other law enforcement agencies take virtually every youth into custody, relying on the probation department to sort out the next steps. A recent development in police handling of juveniles is community/juvenile assessment centers—places at which the police drop off youngsters in instances in which an arrest

Figure 4.1   The Stages of Delinquency Case Processing in the Juvenile Justice System

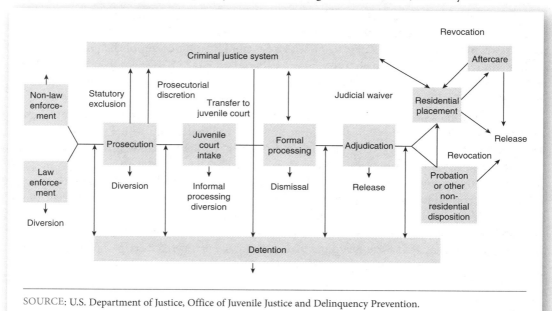

SOURCE: U.S. Department of Justice, Office of Juvenile Justice and Delinquency Prevention.

CASE STUDY: THE CURRENT JUVENILE JUSTICE SYSTEM

Gerald Gault was a 15-year-old who was confined in an Arizona juvenile corrections facility for nine years. It was alleged by the police that he was making an obscene phone call. Mr. Gault claimed that he was calling his mother. There was no direct witness to this alleged crime. There was no formal trial and no transcript of the hearing. The youth was sent by the juvenile court to a juvenile corrections facility where he stayed for almost a decade.

Gault challenged the constitutionality of his court hearing, but the State of Arizona claimed that the supposed benign and treatment-oriented philosophy of the juvenile justice system permitted states to use very informal and flexible procedures. The U.S. Supreme Court disagreed and pointed out the fact that this institution had guards and custodial staff and that Mr. Gault was housed with other young people who had committed far more serious crimes, and that he was at risk to be seriously assaulted.

is not mandatory. An assessment center consists of staff who are multidisciplinary and knowledgeable about community-based treatment options. The primary goal of an assessment center is to perform a treatment assessment for the youth and to quickly move that youngster into more appropriate community services. Police appreciate the assessment center concept because they can drop off the youngsters with responsible adults, it saves them paperwork, and the patrol officers can more quickly return to service.

Because police decisions to arrest or release juveniles are not subject to a great deal of external scrutiny, it is difficult to accurately describe how these decisions are actually made. There is much speculation that race and social class factors unfairly bias law enforcement decisions. Statistics are very hard to collect on such low-visibility decision making. Anecdotal evidence suggests that police are more likely to return youths to their homes if their parents are at home (or reachable by telephone) and if the parents are regarded as reputable by the officers. It is easy to see how class and racial considerations might impact the judgment of a law enforcement officer, particularly one who does not know much about the youngster or his or her community. One of the goals of community policing is to better connect police officers with the residents of the areas that they patrol, setting up the possibility of better decision making as to whether a young person should be arrested.

Researchers have suggested that the demeanor of the youth may play a large factor in whether a formal arrest will be made (Piliavin & Briar, 1964). A belligerent and verbally abusive youth is far more likely to be taken into custody. Moreover, some have observed that teenagers who simply assert their constitutional rights are more likely to be treated in a more formal and harsh manner. The factor of demeanor is sometimes used as a rationale for why it appears that minority youths seem to be arrested in higher proportions than their white counterparts (Piliavin & Briar, 1964). But it should be considered that race differences are pivotal in the way in which different people interpret the behavior of others. A youth who averts his or her eyes from the arresting officer may be showing deferential behavior, although this might be interpreted as a sign of disrespect. Because demeanor is so subjective and situational in its interpretation, it is difficult to comprehend how ethnic, racial, gender, or class differences could not influence this communication process.

There has also been a rising trend in the past 10 years for more youth to be referred to the juvenile justice system by schools. School suspensions and expulsions have more than doubled since the 1970s. Many of these young people are also under probation supervision and so the school actions may trigger additional sanctions that will be imposed by the juvenile court. Some youth advocates have expressed concern that school policies are creating a "school to jail pipeline" (Nelson & Lind, 2015). The rise in expulsions and suspensions followed the popularity of "**zero tolerance**" policies that dictate swift and very negative responses to youth who violate school rules. These zero tolerance policies were marketed as an effective method to keep drugs and guns off of educational campuses. However, there is growing evidence that mandatory suspensions and expulsions are being utilized against students who appear to challenge the authority of teachers. The research is growing that youth of color are vastly disproportionate among those who are harshly punished by school officials. Youth with learning disabilities are more likely to be sanctioned for school behavior. The impacted students are not necessarily accorded protection of their rights, nor are their parents even notified in advance of actions that kick these youth out of schools. Students who are expelled or suspended lose school credits and have a higher likelihood of not graduating (Nelson & Lind, 2015). There are even temporary detentions in schools for some allegedly disruptive young people. The frequency, duration, and nature of these school detentions are rarely subject to public accountability and oversight.

Equally alarming has been the large growth in police who are stationed in schools. From 1997 to 2007, there was an increase of School Resource Officers (SROs) by one third. These officers might be municipal or county law enforcement personnel, employees of the school district, or even private security guards. The qualifications and training of SROs is not standardized or consistent. Almost 92,000 students were arrested in schools during the 2011 to 2012 academic year (Nelson & Lind, 2015).

While the presence of police in schools was justified as a way to protect students from violence, particularly in the aftermath of mass school shootings, most of the arrests were for low-level offenses. For example, the Justice Police Institute found that arrests for "disorderly conduct" were 5 times more likely in schools with SROs, even when one controls for the students' poverty level (Justice Policy Institute, 2011).

Some observers assert that much of the behavior that creates a police response are activities that "make adults mad" but that are not especially dangerous. Moreover, the presence of police in schools may result in situations in which referral to the justice system is used in lieu of less severe conflict resolution at the school level. Put bluntly, the schools may call the cops rather than phone the offending student's parents. Teachers might rely on the SROs to maintain classroom order in lieu of more evidence-based educational interventions. This leads to official police records that are difficult or impossible to expunge. Officials from the juvenile courts have often opposed these in-school law enforcement efforts because they believe that it diverts limited resources away from more serious youthful lawbreaking. As with policies of pushing youth out of school, there is preliminary but substantial evidence that youth of color are more likely to be the targets of arrests by SROs (Justice Policy Institute, 2011).

There are growing demands that the school to jail pipeline be abandoned and that more appropriate disciplinary methods are substituted. Some school districts have completely eliminated zero tolerance policies and others are experimenting with **restorative justice** and peer mediation techniques to reduce school disruptions.

Of those youths who are brought by the police to the juvenile court intake process, slightly more than half of these cases result in the filing of a delinquency petition. About one fifth of the cases that are referred to intake are dismissed outright, and another one quarter receive some informal sanction that might include dismissal, restitution, community service, or a minimal level of probation supervision. For the cases that the intake unit sends forward to the court, about two thirds of these result in formal adjudications. The juvenile court relies on a range of informal sanctions to handle the rest of the cases. Only about 6 in 1,000 referrals to the juvenile court result in the transfer of the case to the criminal court. However, we noted earlier that in some jurisdictions there are statutory provisions permitting the waiver of a juvenile to criminal court without the agreement of the juvenile court (Snyder & Sickmund, 1999).

Since 2001, there has been a drop in the number of juveniles who are arrested by police agencies, especially for violent offenses; the overall number of cases referred to the juvenile court also declined but not as much as the decline in arrests. American juvenile courts handled 3 times as many cases in 2010 compared to 1960; however, this growth in the workload of juvenile courts has slowed since the mid-1990s. The juvenile court continues to process a very large number of young people for minor drug crimes and for violations of probation (Sickmund & Puzzanchera, 2014, pp. 152–153). Whereas the number of males in juvenile court has been going down over the past two decades, the number of females in juvenile court for various offenses has declined more slowly or even increased for both very serious and nonviolent offenses (Sickmund & Puzzanchera, 2014, p. 154).

Most youths who are adjudicated by the juvenile court and who have their petitions sustained by the court are placed on probation or receive lesser penalties. Of all youths that are referred to the juvenile court, about 9% of these cases result in some form of out-of-home placement (Snyder & Sickmund, 1999). These data illustrate how the philosophy of utilizing the least restrictive alternative remains a core principle of the juvenile court.

Because the decision to remove a child from his or her home is such a dramatic (and costly) step, it is worth examining the process by which such determinations are made. The very first placement decision made by the juvenile justice system involves whether a youth should be held in a juvenile detention facility. Approximately 600,000 juveniles are admitted to juvenile detention centers each year. Although many of these youths spend less than 24 hours in detention centers, the average length of stay for all detained youngsters is about 15 days (Krisberg, Noya, Jones, & Wallen, 2001). Federal law requires that the holding of juveniles in police lockups be of very short duration and that youth be moved to an appropriate juvenile facility. Further, this law requires strict "sight and sound" separation of juveniles from adult offenders. The number of cases detained by the juvenile court rose by 50% between 1985 and 1998 and then dropped by 26% between 1998 and 2010. The trend for girls was comparable, although the changes were larger—an increase in detentions of 104% from 1985 to 1998. There was a 43% decline in girls sent to detention centers from 1998 to 2010. Generally, boys were more likely to be detained than girls. Older adolescents were more likely to be detained than those under age 15, and youth of color, especially African American young people, were more likely to be detained than white youngsters (Sickmund & Puzzanchera, 2014, pp. 162–163). These disparities in the treatment of females and minority youngsters are discussed in more detail in later chapters.

Initial detention decisions are typically made by intake workers. While there are some states that define general criteria for detention, there is often very wide latitude given to the juvenile justice system. Unlike the adult court system, juveniles may be detained for their own protection, as well as to maintain public safety. Juveniles do not have the right to post bail in most jurisdictions. Many states bar the confinement of status offenders in secure detention facilities, although there are often statutory loopholes that allow for the incarceration of youngsters who have committed status offenses. These exceptions to the ban against confining status offenders can include youths who have violated a valid court order or a youngster who is already on probation.

In recent years, several juvenile justice systems have employed structured **risk assessment** instruments (RAIs) to assist in making detention intake decisions. These RAIs employ a limited set of variables that have been demonstrated via research to determine the risk of committing another crime before the case is adjudicated or the risk of the youth failing to make scheduled court hearings. Similar instruments have been utilized in making pretrial confinement decisions in the criminal court system since the 1950s. The primary focus of the RAI is the protection of public safety. The intake worker scores a specific case and generally follows the presumptive custody or release decision that is suggested by the RAI. Intake workers can override the presumptive decision of the instrument, but many departments require that a supervisor review and approve this override.

The use of more structured and objective detention intake instruments has led to significant declines in admissions in a number of locations (Krisberg et al., 2001; Schwartz & Barton, 1994). Reducing detention admissions through the use of well-designed RAIs does not appear to endanger public safety, nor does it lead to increased failure-to-appear rates (Krisberg et al., 2001; Steinhart, 1994). Moreover, it appears that the use of RAIs can reduce the amount of racial, gender, and class bias that often enters into more subjective detention intake systems (Wiebush, Baird, Krisberg, & Onek, 1995).

In most locales, the alternatives to secure detention are fairly limited. Intake workers either decide to hold the youth pending a court review or they generally release the child to a legal guardian. More recently, a range of additional alternatives to detention have been attempted. For example, some jurisdictions have employed electronic monitoring devices that allow the probation department to determine if the minor is at home during specified hours. Other locales have funded day reporting programs, in which the youth lives at home but is expected to show up daily at a program, usually after school. Detention alternatives may also involve intensive home-based visits by probation staff or by a designated community-based agency. For youths who do not pose a threat to public safety but need a safe place to live temporarily, some juvenile courts have established nonsecure shelters where the youngster can stay until the court proceedings are completed. Research has demonstrated that properly implemented alternatives to detention are considerably less costly than secure confinement, and youths placed in alternatives make their court hearings and do not commit many new offenses while they are living at home (or in the community) awaiting the final disposition of their cases. In fact, well-designed alternatives can actually reduce failures to appear and pretrial crimes because youths in these programs are in greater contact with court personnel in the preadjudication period. Youths who are initially detained are often released before their court hearings are completed and may reoffend, even though they had been initially incarcerated (Krisberg et al., 2001; Schwartz & Barton, 1994).

Reducing the number of youngsters who are inappropriately incarcerated during the preadjudication period is desirable for a number of reasons. As noted earlier, the decision to detain is expensive and often creates crises within families trying to reconcile differences and reunite. Minors who are detained pretrial are more likely to be placed out of their homes after their cases are disposed (Schwartz & Barton, 1994). While this correlation may be a function of the severity of the offense or the past delinquent history of the child, there is some evidence that the initial decision to detain is a self-fulfilling prophesy in that court officials assume that detained youths must be more serious threats to public safety.

Too often, detention centers do not have effective custody **classification systems**, so that minor offenders may share the same living unit with more aggressive and criminally sophisticated youths. Many detention centers are dominated by gang conflicts and violence. Little in the way of treatment or education takes place in most secure detention centers. Youngsters in detention do "dead time." There are few medical, behavioral, or mental health services available to children in detention. Youths with serious health issues often deteriorate in these stark settings (Hartney, Wordes, & Krisberg, 2002). Most progressive thinkers in the juvenile justice system view detention as a necessary evil to be limited as much as possible.

As problematic as the conditions in juvenile detention centers are, the practice of some jurisdictions of putting minors in adult jails is even more dangerous. Historically, minors placed in adult jails have experienced much higher rates of violence and sexual exploitation than youths placed in specialized youth facilities. The suicide rate for juveniles in jails is much higher than youths in juvenile detention centers.

The passage of the federal Juvenile Justice and Delinquency Prevention Act of 1974 (JJDPA) was, in part, motivated by an effort to end the abuses suffered by children who were held in adult jails and police lockups. At the time of the enactment of the JJDPA, it was estimated that nearly 500,000 minors entered adult jails each year. There were 100,000 minors admitted to jails in California alone in the early 1980s. Initially, the JJDPA sought to impose strict separation between adult and juvenile inmates. Later amendments to the law called for the complete removal of children from adult facilities. Congress allowed each state to apply its own legal definition of a juvenile. A more conservative Congress voided the jail removal standard and returned to the goal of separating adult and juvenile jail inmates, but the overall trend has been to provide more loopholes for jurisdictions that wish to continue this practice.

The national reform to reduce the jailing of children met with substantial success. By the late 1980s, juvenile admissions to jails had declined to roughly 65,000 per year (Austin, Krisberg, & DeComo, 1995). The more conservative juvenile justice policies of the 1990s led to a 35% increase in children in jails between 1994 and 1997 (Snyder & Sickmund, 1999). In 1997, of the approximately 9,100 youth under age 18 who were jail inmates, three quarters were awaiting trials or were convicted in criminal courts (Snyder & Sickmund, 1999).

The jailing of children has historically been a problem for rural jurisdictions and those locales where large distances prevent law enforcement officials from quickly transporting youngsters to juvenile facilities. However, very punitive attitudes among some judges also contributed to the dangerous practice of holding minors in adult facilities. Reformers employed a range of tactics to reduce the jailing of children. Some jurisdictions built new juvenile facilities, while others provided funding for community-based shelter facilities. A major bipartisan legislative campaign in California made it illegal to jail minors for more than 6 hours and then only for the purpose of moving them to a more appropriate setting (Steinhart, 1988). The practice of jailing children in California virtually disappeared after the enactment of this law.

Sending minors to adult prisons was not part of the national jail reform campaign. There are no federal laws prohibiting states from placing very young adolescents in state prisons if they have been convicted in criminal courts. Indeed, this practice has become much more common over the past 10 years (Torbet et al., 1996).

As we have already observed, no one knows for sure how many juveniles are transferred to adult courts. Hamparian (1982) estimates that in 1978 there were more than 9,000 youths transferred to criminal courts by juvenile courts, another 2,000 cases moved to adult courts by prosecutors, and more than 1,300 minors prosecuted as adults due to statutes that excluded certain offenses from juvenile court jurisdiction. Later estimates from a variety of sources suggest that as many as 200,000 youths under age 18 are tried in adult courts (Krisberg, 1997).

The National Correctional Reporting Program found that there were nearly 7,400 persons under age 18 who entered prisons in 1992 from the 37 states participating in that national data system (Strom, 2000). These minors represented about 1% of all prison admissions from these states. These youngsters were predominantly African American males (58%) and were typically 17 years old at the time of their prison admission. About 47% of these youths were convicted of violent crimes, and 53% were charged with drug offenses or property crimes (Krisberg, 1997). Youths under age 18 who enter prison receive longer sentences and actually serve 3 times as long in prison compared to adults convicted of similar offenses. Howell (1997) notes that juveniles in prisons are more likely to get involved in disciplinary problems that extend their prison stays, and they serve a very large proportion of their sentences in prison administrative segregation units.

Although there are many studies of the process through which juvenile courts select cases to transfer to criminal courts, virtually nothing is known about how prosecutors choose cases to file directly in criminal courts or how laws restricting the jurisdiction of the juvenile court are actually enforced. Howell (2003) found 38 studies of judicial waiver that sought to account for the decision-making process. He found that legal factors such as severity of the crime and prior record influenced the waiver decision. Howell also found that race, the age of the offender, victim–offender relationships, and local juvenile court practices exerted strong effects on the decision to transfer youngsters to adult courts. For example, compared to white minors who were charged with the same offenses, African American youths had significantly higher rates of transfer for drug crimes and violent offenses (Snyder & Sickmund, 1999). Frazier (1991) finds clear evidence of racial differences in the use of prosecutorial discretion to file juvenile cases in adult courts in Florida. Eigen (1972) reports that African American teenagers who murdered white victims were much more likely to be tried as adults than as juveniles in Philadelphia. Singer and McDowall (1988) show how New York State's tough laws on trying juveniles as adults were selectively enforced in a number of different New York counties. Krisberg (2003) illustrates how racial bias and inadequate legal representation can propel minority youngsters into the adult criminal court system.

The decision to remove juvenile offenders from the juvenile justice system constitutes a very dramatic change in a young person's legal status. We noted earlier that juveniles in prisons are less likely to receive treatment services, they are more likely to be victimized by older inmates, and they spend large parts of their incarceration time in solitary confinement units (Forst, Fagan, & Vivona, 1989; Howell, 2003). Transfer decisions, however, have even more stark consequences for young offenders. For example, no juvenile court permits the imposition of capital punishment, but youths who are 16 and older and are tried in criminal courts may be put to death. In 1998, there were 76 death-row inmates who committed their crimes before the age of 18. Forty percent of these cases were in just two states—Texas and Florida. The United States is virtually alone among countries in the developed world that permitted juveniles to be put to death (Streib, 1987). In successive cases such as *Thompson v. Oklahoma* and *Roper v. Simmons*, the U.S. Supreme Court has effectively ended this discredited practice.

In the last few years, we have witnessed ever younger adolescents being tried as adults and sent to state prisons. The tender age of these prison-bound youths has shocked the conscience of some, but there are the beginnings of a national reform movement to stop this policy (Amnesty International, 1998; Daugherty, 2015). There has been some limited progress to end this practice, but federal and state laws continue to support the wider use of transfer procedures.

SUMMARY

There is major variation among the states and even within states with regard to the structure and operation of the juvenile justice system. Previously, my colleagues and I referred to this situation as "justice by geography." Most states set the upper age boundary of original jurisdiction of the juvenile court at 18, but this age limit can be as low as 16 years old in some states and as high as 25 years old in others. Juvenile justice systems have become more like criminal justice systems over the past years, but there remain major differences between the two. Unlike the adult system, the juvenile justice system is guided by a philosophy of rehabilitation and child protection. Juvenile justice systems are more focused on preventive practices and actively try to divert young people from the formal court process. The criminal justice system has a lesser emphasis on diversion and principally pursues prevention through the notion that its penalties serve as a deterrent to potential offenders.

Most cases enter the juvenile justice system through police referrals, but schools, social agencies, and parents also bring cases to juvenile court. Most communities operate juvenile intake units that evaluate individual cases and attempt to find alternative methods of resolving the immediate issue. Of youths brought to intake units, slightly more than one half go on to formal court hearings. Discretionary decision making is a core part of how the juvenile justice system operates. Much of the research on juvenile justice has focused on the determinants of these decisions.

Increasingly, states have enacted more ways to transfer young people to the criminal courts. This is a very controversial trend, and there is a great deal of research suggesting that trying very young people in criminal courts is bad public policy. For example, no juvenile court can impose the death penalty, but juveniles transferred to adult courts can receive capital punishment.

REVIEW QUESTIONS

1. Compare and contrast ways in which the criminal justice and juvenile justice systems work.

2. What are the main findings of research on how decisions are made in various stages of the juvenile justice process?

3. What are the presumed benefits and problems of transferring youths from the juvenile justice system to the adult system?

INTERNET RESOURCES

The National Council of Juvenile and Family Court Judges is the preeminent source of information on the operation of the juvenile court.
http://www.ncjfcj.org

The Vera Institute of Justice is a leading resource on research and policy on the adult and juvenile justice system.
http://www.vera.org

The Annie E. Casey Foundation is the leading philanthropy that focuses on juvenile justice issues.
http://www.aecf.org

Sage Publications is the leading publisher of texts and journals on crime and delinquency.
https://us.sagepub.com

Chapter 5

JUVENILE JUSTICE AND THE AMERICAN DILEMMA

At the dawn of the 20th century, the great American sociologist and social activist W. E. B. DuBois proclaimed that the "problem of the color line" would be the central issue of that century (DuBois, 1903). As we are now in the early decades of the 21st century, his prophetic words still ring true. The juvenile justice system is plagued by racial disparities and inequities that challenge its legitimacy. Despite a dominant philosophy that seeks to be "race neutral," the decision-making processes of the juvenile justice system are anything but neutral for children of color. In this chapter, the extent of this grave problem is examined, explanations for its existence are explored, and potential remedies are considered.

In 1992, the federal Juvenile Justice and Delinquency Prevention Act (JJDPA) was amended by Congress to require that all participating states wishing to receive funding from the Office of Juvenile Justice and Delinquency Prevention (OJJDP) be required to examine the extent of disproportionate minority confinement (DMC) in their juvenile justice systems. Besides conducting in-depth studies of the problem, participating states were required to make "good faith efforts" to reduce DMC where appropriate. Since the late 1990s, there have been annual attempts from some members of Congress, most notably Senator Orrin Hatch of Utah, to remove the DMC mandate from the federal justice program. Although these efforts have weakened the mandate and given states great flexibility in reporting, Congress has not yet eliminated the requirement for states to examine racial disparities in their juvenile justice systems.

Krisberg and his colleagues (1987) utilized data from the national Children in Custody survey to document the disproportionate presence of African American, Native American, and Latino youths in juvenile correctional facilities. Further, they found that minority youths were more likely to be housed in more secure facilities and to be held in places that were chronically overcrowded. Moreover, Krisberg and his associates noted that reform efforts in the 1970s that were intended to deinstitutionalize youths, removing them from secure facilities to community settings, had primarily resulted in white youth incarceration rates going down and far less of a decline in the confinement of minority youngsters. As the public policy agenda embraced tougher penalties and longer stays in custody in the 1980s, minority youths bore the overwhelming brunt of these "get tough" policies. Throughout the next several decades, racial disparities continued to worsen.

Before examining some of the key recent data on DMC, it is important to clarify some basic terminology. Although it is common to use the word *race* to describe certain groups, it must be noted that mainstream scientists regard race as a meaningless biological category. Research on human genetics has consistently shown that human genetic diversity is very limited in comparison to other living organisms. Increasingly, scientists believe that observable human differences are attributable more to geographic communities (i.e., people who have lived in close proximity for generations) than to genetic differences. It is highly questionable science to use race as an explanatory variable in and of itself, although several conservative, media-promoted criminologists such as Charles Murray and James Q. Wilson have consistently produced flawed research using scientifically questionable data in this area (Herrnstein & Murray, 1994; Wilson & Herrnstein, 1985). Further, the extensive sexual contacts among racial groups, which is often publicly denied, makes pure racial distinctions moot. Most Americans have interracial heritages, and there are no scientifically valid rules by which an individual should be assigned to one group or another. Race is, first and foremost, a social construct and a powerful life-defining category. Race-conscious societies often have intricate rules, because assignment to one or another racial group *does* have profound social and even legal consequences. For years, many southern states such as Louisiana made very fine distinctions based on the presumed percentage of one's blood that came from different racial groups. For some, even "one drop of blood" defined a person as belonging to a particular racial category. There is a rich literary and folk tradition about people who attempted to "pass" as members of another group (for example, the classic American musical *Showboat* or the classic movie starring Lana Turner, *Imitation of Life*). Racial categories are fluid and change as society changes. At the turn of the 20th century, Jewish and Italian immigrants were regarded as nonwhite. During certain historical periods, being part Native American was a badge of honor, but at other times this racial background was a **stigma**.

Related to the concept of race is that of *ethnicity*. Ethnicity refers to a group of people who generally share a common culture. The commonality of an ethnic group might include history, values, aesthetics, family and community traditions, and language. Most of the groups who are referred to as ethnic groups may share some of these common social and cultural elements, but more often than not, there is notable diversity within these groups. For example, the category *Latino* may refer to people with a heritage from North American, South American, Central American, or Caribbean nations. They practice several different religions and speak English, Spanish, or Portuguese, among other languages. Similarly, there is enormous cultural diversity in the African American community, whose antecedents may have originated in Africa but who lived in virtually every nation on earth. Even for those born in one nation, regional differences or rural-versus-urban experiences profoundly affect a person's ethnic identity. There is a serious question as to whether a person can self-select an ethnic identity, such as the white teenager who totally embraces the culture of urban African Americans, or the Eastern European immigrant who attempts to embrace the culture of white Anglo-Saxon Protestants. As with race, the concept of ethnicity is a social construct and is most meaningful in terms of how others react and respond to an individual, as well as the ethnic identity that one projects to the outside world.

Both race and ethnicity are further complicated by differences in gender and social class. Often, in discussions about DMC, people will ask whether the observed differences are really a function of economics or other indicators of social prestige. As some would observe in the trial of famous football star O. J. Simpson, the true color of justice may be green. There is no easy answer to this question. While justice agencies routinely collect data on the presumed race or ethnicity of defendants, they rarely compile data on socioeconomic status. Only more detailed data collection sometimes teases out these relationships. Further, there is almost as much ambiguity about the social meaning of class as there is about race and ethnicity. How gender differences further complicate matters is discussed in the next chapter.

If you are feeling confused and getting a mild headache after considering these complexities, you are probably getting the right messages. Terms such as *race*, *ethnicity*, and *social class* are used imprecisely and sometimes interchangeably. This is a big problem that is embedded in the existing data and research. There is no simple solution to this conceptual quagmire except to recognize that it exists and frustrates both good research and sound public policy discussions on this topic. One thing is clear—all of these categories are socially constructed, and therefore, they can be made and unmade by people.

Disproportionate Minority Representation and the Juvenile Justice Process

The federal DMC mandate has produced a rich mix of data and analytic insights on the problems of race and juvenile justice across many states. OJJDP called for analyses of the entire juvenile justice process—the ways in which decisions at each level of the process might culminate in DMC. A national summary report of dozens of these state-specific studies that was authored by Hamparian and Leiber (1997) found that overrepresentation of minority youths increased at every stage of the juvenile justice process. Although the research indicates that the largest gap between white youths and children of color occurred at the front end of juvenile justice—at the point of arrest, court intake, and detention—the data also reveal that the initial disparities got worse at later stages in the juvenile justice process. This finding has led some observers to talk about a "cumulative disadvantage" for minority youths in the justice system (Leonard, Pope, & Feyerherm, 1995; Males & Macallair, 2000).

The most comprehensive description of this cumulative disadvantage was produced by Poe-Yamagata and Jones (2000) in their path-breaking report titled *And Justice for Some*. In this study, the authors incorporate data on DMC from a very wide range of federal data sources on juvenile justice as well as on youths in the adult corrections system. The report traces the disproportionate representation of minority children from arrest through the dispositional or sentencing process.

The vast majority of youths first encounter the juvenile justice system through contacts with the police. Law enforcement responses to alleged instances of juvenile misconduct can cover a broad range of actions, including verbal reprimands and informal case handling to arrest and detention. The first decision that is made is whether to make an arrest, and police are given fairly wide latitude in making this decision. Typically, police divert or dismiss about 25% of the youths that they arrest; about two thirds of these cases are referred to the juvenile court, and about 10% are sent to the criminal court system or are referred to other community agencies. Studies of police practices by Wilson (1968), Cicourel (1968), and Emerson (1969) have shown that there are very wide differences among police agencies in the ways in which these discretionary decisions are made at the front end of the juvenile justice system.

There were roughly 1.6 million persons under the age of 18 who were arrested in 2010, with 66% of these arrests involving white youngsters. However, African American youths are disproportionately represented, comprising 31% of those arrested compared with being 17% of the total population of youths under age 18 (Sickmund & Puzzanchera, 2014). Table 5.1 breaks down these data for several ethnic groups and offense categories. The reader should note the inability of current federal statistics to provide information on Hispanic or Latino youths. For most data sources, Hispanic youths are combined with whites. Also, we will later discuss how the aggregate crime statistics mask the very high rates of juvenile justice involvement of youths from certain Pacific Islands and first- and second-generation migrants from Southeast Asian countries.

In 2010, African American youths made up 17% of all those under the age of 18 in the general population; whites were 76%, Asians were 5%, and Native Americans were 1% of the youth population ages 10 to 17 (Sickmund & Puzzanchera, 2014). Compared to their percentage of the youth population, African American youth were also disproportionally arrested for weapons offenses, prostitution, gambling, disorderly conduct, and curfew and loitering violations. When compared to their size of the general youth population, the only offense categories in which white youth are disproportionately arrested are liquor-law violations, driving under the influence, and public drunkenness. Because these latter offenses usually occur in conjunction with traffic violations, the greater likelihood of white youths having access to cars may explain these findings.

It is worth noting that as the number of juveniles who were arrested declined over the past decade, the racial disparities among those arrested grew. For example, African American youth constituted 28% of those arrested in 2004 and 31% of those youth aged 10 to 17 in 2010; whites were 70% of those arrested in 2004 and 66% of those arrested in 2014. The relative proportions of arrested Asian American or Native American youth did not change during that period. The

Table 5.1   Racial Proportions of Arrested Youth Under Age 18, 2010

Most Serious Offense Charged	Estimated No. of Juvenile Arrests	Percent of Juvenile Total Arrests			
		White	African American	Native American	Asian
Total Arrests	1,642.500	66	**31**	1	2
Violent Crime Index	75,890	**47**	**51**	1	1
Murder	1,100	43	56	0	1
Rape	2,900	63	36	1	1
Robbery	27,200	31	67	0	1
Aggravated assault	44,800	56	41	1	1
Property Crime Index	366,600	**64**	**33**	1	2
Burglary	65,200	62	36	1	1
Larceny/theft	281,100	65	32	1	2
Motor vehicle theft	15,800	55	42	1	1
Arson	4,600	75	18	1	1

NOTE: Detail may not add to total due to rounding.
We have reported on the offenses included in the FBI crime index because these offenses are the most likely to be consistently recorded by police agencies. The data do not disaggregate Latino youth by race, but most Latino youth are identified by law enforcement officials as white.
SOURCE: Adapted from Sickmund & Puzzanchera ed, (2014). U.S. Department of Justice, Office of Juvenile Justice and Delinquency Prevention.

news was that somewhat fewer African American youth were arrested, but the racial differential of arrested African Americans compared to white youth worsened.

Research by David Huizinga and Delbert Elliott (1987) suggests that the overrepresentation of minority youth, and especially African American youngsters, cannot be explained by a higher level of offending by those groups. Using self-report questionnaires as part of the National Youth Survey, Elliott and others were able to construct national estimates of the age, race, gender, and social class distribution of offending by juveniles. There has been some concern expressed that African American youths underestimate the extent of their criminal behavior in self-report surveys. However, Elliott, Huizinga, and Ageton (1985) conclude that this alleged underreporting is not statistically significant. In general, African American youths report a slightly higher level of delinquency, but these differences were not statistically significant. Further, African American youths did report somewhat higher proportions involved in the more serious offenses, and their frequency of offending was greater than for white youths, but these modest differences could not explain the wide discrepancies among the groups in terms of arrests or other indices of system processing. Huizinga and Elliott (1987) also report that the delinquent acts that African American youths admit to are far more likely to result in arrests than if those same offenses are claimed by white youths. Their conclusion may reflect, among other factors, the density of deployment of police in minority communities compared to white areas. Police rarely go into middle- and upper-class neighborhoods looking for teen crime violations. Most police contact with juveniles takes place in public settings, including schools, playgrounds, and city streets. Also, it appears that police use informal adjustments and voluntary referrals to a greater extent for white youths. Police may perceive (sometimes correctly) that white communities possess more prevention and treatment resources to deal with offending youths.

There is some reason to believe that the legacy of police violence against persons of color and the history of excessive use of force against minority citizens exert a profound influence on the ways in which minority youths and adults interact with the police. Some observers have tried to explain the more formal and restrictive response of predominantly white law enforcement officers to minority youths in terms of the concept of demeanor. Minority youths are allegedly more hostile, more confrontational, and more verbally challenging to white authority figures than their white counterparts. There is scant empirical evidence to back these claims, but it is not hard to understand how historical racial patterns influence the perceptions of demeanor in tense situations (Piliavin & Briar, 1964).

The disparity among ethnic groups seen at the arrest stage also is reflected in the composition of the juveniles that come to juvenile court. African American youths comprise 33% of those sent to the juvenile court—almost twice their proportion in the general youth population. There is some evidence suggesting that court intake workers may further exacerbate the racial differentials seen at the arrest or court intake levels. At the point of the decision to detain, things get even worse for minority youths. For example, white youngsters comprise 64% of those referred to the juvenile court but only 19% of those detained. African Americans make up 33% of the court referrals but 25% of those detained (Sickmund & Puzzanchera, 2014). The higher likelihood that African American youths will be detained is true for both violent offenses and property crime. However, the greatest differential among the groups occurs with drug offenses. Black youths make up about one third of those referred to the juvenile court for drug offenses, but they constitute 55% of those who are detained for drug crimes. Put differently, an African American youngster who is brought to court for a drug offense is almost twice as likely to be detained as compared with their white counterpart. Native American and Asian American youngsters are also detained at a disproportionate rate compared to white youth, even though they represent a much smaller share of the caseloads of the juvenile court.

Minority youngsters also spend more time in detention than white youths. These disparities hold even when one controls for whether the youth had prior referrals to the juvenile court before the current charge (National Council on Crime and Delinquency [NCCD], 2001).

Detention is a dramatic loss of liberty and may begin a further downward spiral toward deeper penetration in the justice system. Being more likely to be detained also means that the youth is more likely to be placed out of home or incarcerated upon adjudication (Poe-Yamagata et al., 2007). African American youths are more likely held in overcrowded and substandard urban detention centers in which treatment and educational resources are virtually nonexistent.

Current statistical data on the juvenile court suggest that there is a smaller disadvantage for minority youths as they move through the adjudication stage compared to earlier steps in the juvenile justice process. Still, African American teens are more likely to receive a formal adjudication for delinquency for virtually every offense category. This disparity is the most pronounced for drug offenses. In 1997, roughly 78% of drug cases involving African American youths resulted in a formal petition of delinquency, compared to 56% of cases involving white youths (Snyder, 1999; Snyder, Finnegan, Stahl, & Poole, 1999). Here again, we see that the **War on Drugs** has had an extremely adverse impact on young people of color.

As noted earlier, there is an increasing tendency in law and practice to transfer cases from the juvenile court to the adult criminal courts. This decision can be made mandatory based on state statutes or may result from prosecutors having the discretion to send juvenile defendants directly to the adult courts. Roughly 8,400 cases per year that are referred to juvenile courts are transferred to the criminal courts by virtue of a hearing in the juvenile court. As with other juvenile justice decision points, white youths are underrepresented in the population of those sent to the criminal courts, whereas African American young people are overrepresented in the referred population. For example, African American teens comprise about 34% of those referred to the juvenile court, but they are 46% of those referred, or waived, to the criminal court. The racial disparity is especially evident in the way in which the juvenile court handles crimes against persons and drug crimes (Poe-Yamagata et al., 2007).

A study by the Pretrial Services Resource Center based on 18 urban jurisdictions showed that minority youths sent to the criminal courts suffered a similar cumulative disadvantage to those retained in the juvenile court. During the first 6 months of 1998, a stunning 82% of all minors sent to adult courts in these 18 jurisdictions were youths of color (Juszkiewicz, 2000). In six of the studied jurisdictions, minority youths comprised more than 90% of those children tried in criminal courts. For example, in Jefferson County, Alabama, African American youths comprised roughly 30% of those arrested for felonies, but they made up nearly 80% of those tried in criminal courts.

The most glaring race disparities found in the Pretrial Services Resource Center study were for drug crimes and for public order offenses. Adult felony drug charges were filed against African American youths at a rate 5 times that of white youths and 3 times that of Latino youths (Juszkiewicz, 2000).

In most instances, the adult court charges resulted from the actions of prosecutors, as opposed to waiver decisions made in the juvenile court. Depending on the locale, youths tried in the criminal courts were confined pretrial in either jails or juvenile facilities. Large numbers of these youths who were tried in criminal courts were not convicted. For example, less than half (43%)

of the African American youths who were charged with felonies were actually convicted, compared with 28% of Latino youths and 24% of whites. About 20% of the African American youths were transferred back to the juvenile court. These data are somewhat ambiguous in their interpretation. One might conclude that criminal courts are more lenient toward minority youngsters (a highly unlikely hypothesis); alternatively, it might be argued that prosecutors overcharge these cases and bring weaker cases to court when these involve African American teenagers. For those convicted in criminal courts, African American youths were more likely to receive a sentence of imprisonment for virtually every offense category. Of those sentenced to incarceration, African American youths received longer sentences than other youths (Juszkiewicz, 2000).

If cases remain in the juvenile court system through adjudication, the familiar pattern of racial disparity continues. African American youths are more likely than white youths to be placed out of their homes for every major offense category. White youths are more likely to be placed on probation. As observed above, drug offenses seem to produce the largest differences in outcomes among the racial groups. For white youths adjudicated as delinquents for drug crimes, 61% receive probation; for African American youths, 49% receive probation.

YOUTHS OF COLOR IN CONFINEMENT: THE NATIONAL PICTURE AND STATE DIFFERENCES

The national-level data on juvenile justice processing shed considerable light on the tragic overrepresentation of minority children in confinement. However, it is important to remember that the localized variations in juvenile justice laws and practices leading to this problem are even more severe in certain jurisdictions. OJJDP has promulgated a very simple measure of disproportionate minority confinement. The calculation compares the proportion of a given ethnic group in the confined population to that same population in the general youth population. A ratio of 1.0 would suggest that minority youths are represented in the confinement population at roughly the same level as that of the general youth population. If this same ratio is less than 1.0, this would signify underrepresentation. The more the ratio exceeds 1.0, the larger the extent of overrepresentation of minorities in custody. Let us examine the problem of DMC in a range of custody settings.

An obvious place to start is juvenile detention facilities. Detention is the entry point for the confinement system, and these lockups are the most frequent form of secure custody experienced by young people. A summary study of DMC reports submitted to OJJDP found that all but one state reported disproportionate minority confinement in juvenile detention centers (Hamparian & Leiber, 1997). This report found an average index of minority overrepresentation of 2.8—meaning that minority youths were confined in detention centers at a rate that was 280% higher than their proportion of the general youth population. Only Vermont reported an underrepresentation of minority youth in detention, or a ratio of 0.7. By contrast, Iowa reported the highest overall DMC ratio of 7.9.

Table 5.2 presents data on the overrepresentation in detention of African American youngsters across a number of states. In every state except Vermont, the residential rate for African Americans exceeded the rate of white youngsters. Topping this list were Nebraska, Oregon, Wisconsin, and Kansas

The Latino residential placement rate is generally lower than for African Americans but higher than white youth. The U.S. placement rate per 100,000 for Hispanic youth was 228 compared to

Table 5.2   The Rate per 100,000 of African American Youth in Residential Placements, 2010

U.S. total	606	Delaware	705	Massachusetts	404	Oregon	1,213
Alabama	393	District of Columbia	501	Michigan	627	South Carolina	451
Alaska	643	Florida	652	Minnesota	673	Tennessee	294
Arizona	334	Illinois	478	Nebraska	1,715	Texas	530
Arkansas	535	Indiana	719	New Jersey	540	Vermont	0
California	988	Kansas	1,040	New Mexico	651	Virginia	584
Colorado	1,201	Louisiana	473	New York	539	Washington	624
Connecticut	361	Maryland	322	North Carolina	249	Wisconsin	1,064

NOTE: The rates were calculated by dividing the number of African American youth in residential placement on February 24, 2010, by the number of African American youth in the 2010 juvenile population as measured by the U.S. Census Adapted from Sickmund & Puzzanchera ed, (2014). U.S. Department of Justice, Office of Juvenile Justice and Delinquency Prevention

128 for white youth and 606 for African American young people. States with very high rates of Latino youth in residential placements include California, Nebraska, South Dakota, Oregon, Pennsylvania, Vermont, West Virginia, and Wyoming. The lowest Latino residential placement rates were found in many southern states such as Florida, Louisiana, Maryland, and Mississippi. But these state data on Latino youths in custody must be reviewed with some skepticism because jurisdictions utilize very different criteria in labeling youngsters as Latino, and there is far less care taken in data collection with respect to Latino youngsters (Villarruel & Walker, 2002).

The residential custody rate for Asian American and Pacific Islander youth was 47 per 100,000 youths and was 369 per 100,000 for Native American youngsters (Sickmund & Puzzanchera, 2014).

African American and Latino youths were more likely to be placed out of home, not only for very serious offenses but also for virtually every offense group. Native American youth were more than one third as likely as white youths to be confined for serious crimes against persons or property and were more likely to be confined for public order crimes such as vagrancy, vandalism, and curfew violations. African American youth were twice as likely as white youth to be in residential placements and for drug crimes. Minority youngsters also were more likely than white youths to be locked up for **technical violations** of probation (Poe-Yamagata et al., 2007).

There are other data suggesting that minority youths are primarily confined in public correctional programs and white youths are sent in greater numbers to private facilities (Poe-Yamagata et al., 2007). Private facilities are smaller, are not likely to be locked facilities, have a greater investment in treatment and educational programs, and are rarely overcrowded. Public juvenile corrections facilities are chronically understaffed and under budgeted, and often have very poor conditions of confinement.

A study based on data on admissions to state juvenile correctional facilities shows that African American and Latino youngsters had much higher admissions rates—and this held true both for youths with no previous experiences in state custody and for chronic offenders (see Figures 5.1 and 5.2; Austin, Krisberg, & DeComo, 1995). These same data reveal that both Latino and

African American youths had longer stays in custody facilities compared to whites, even when one controlled for the severity of their commitment offenses.

Figure 5.1   1993 Admission Rates* of Juveniles to State Public Facilities

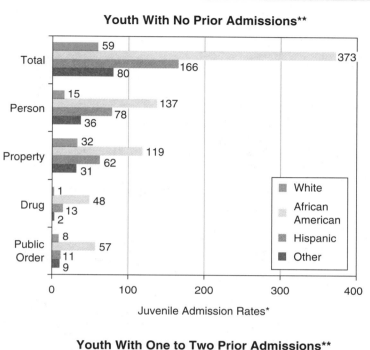

Youth With No Prior Admissions**

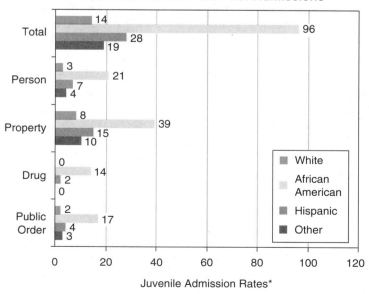

Youth With One to Two Prior Admissions**

(Continued)

Figure 5.1 (Continued)

Juvenile Offender Population by Facility Operation, 2014

Facility operation	Number of facilities	Percent of facilities	Number of juvenile offenders	Percent of juvenile offenders
Total	1,852	100%	50,821	100%
Public	1,008	54%	36,110	71%
State	390	21%	17,200	34%
Local	618	33%	18,910	37%
Private	844	46%	14,711	29%

Note: Detail may not add to total because of rounding

1993 Admission Rates* of Juveniles to State Public Facilities
* Rates are calculated per 100,000 youth age 10 to the upper age of juvenile court jurisdiction in each state.
States include AK, AZ, AR, CA, DE, GA, ID, IL, IN, IA, KY, LA, ME, MD, MA, MN, MS, MO, NE, NH, NJ, NY, ND, OH, OK, OR, SC, SD, TN, TX, UT, VT, VA, WV, and WI.
NOTE: Persons of Hispanic origin may be of any race. White and African American categories do not include youth of Hispanic origin.
Totals contain offenses not shown.
SOURCE: National Council on Crime and Delinquency, *And Justice for Some*, Poe-Yamagata and Jones (2007).

YOUTHS IN PRISON

There were approximately 4,100 persons under the age of 18 who were admitted to state prisons in 1997 (Poe-Yamagata & Jones, 2000). Sentenced by criminal courts and regarded as adults for the purposes of their punishment, these young people are not protected by the provisions of the federal JJDPA, which mandates sight and sound separation between juveniles and adult inmates. Almost three quarters (73%) of these adolescents sent to prison were minority youths. African American young people alone accounted for more than one half (58%) of these prisoners younger than age 18; Latino youth made up 10% of this group. The African American teenage population in prisons grew from 53% of the total to 58% from 1985 to 2002. During this same period, the proportion of the population of white prisoners younger than age 18 declined. In 2002, the prison admission rate of African American males was 84 per 100,000 compared with 9 per 100,000 for white youth, 14 per 100,000 for Latino young people, and 13 per 100,000 for Native American youngsters. For female juveniles, the prison admission rate per 100,000 was 3 for African American girls as compared with rates of 1 per 100,000 for white and Latino girls and 2 per 100,000 for Native American girls.

Drug crimes have played a major role in the increasing presence of African American teenagers in prison. In 1985, the proportion of teenagers in prison for drug crimes was roughly 2% for both African Americans and whites. By 1997, this proportion of drug offenders had grown to 5% for whites but was 15% for African Americans (Poe-Yamagata et al., 2007).

One study revealed that in Alabama, Georgia, Mississippi, Virginia, South Carolina, and North Carolina, African Americans were more than three quarters of those under age 18 that were sent to prisons. In Utah and Colorado, Latino youths made up almost one half of prison admissions of minors. In South Dakota and North Dakota, Native American youngsters constituted 45% and 40%, respectively, of new underage prisoners (Perkins, 1994). There is little

Figure 5.2   1993 Youth Mean Lengths of Stay in State* Public Facilities

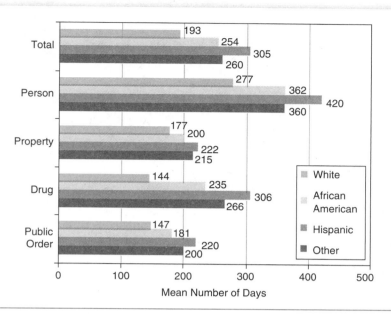

* States include AK, AZ, AR, CA, DE, GA, ID, IL, IN, IA, KY, LA, ME, MD, MA, MN, MS, MO, NE, NH, NJ, NY, ND, OH, OK, OR, SC, SD, TN, TX, UT, VT, VA, WV, and WI.

NOTE: Persons of Hispanic origin may be of any race. White and African American categories do not include youth of Hispanic origin.

Totals contain offenses not shown.

SOURCE: Poe-Yamagata and Jones (2000).

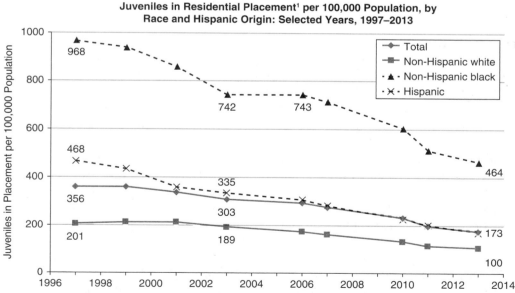

Juveniles in Residential Placement[1] per 100,000 Population, by Race and Hispanic Origin: Selected Years, 1997–2013

[1]The Census of Juveniles in Residential Placement collects data from all juvenile residential custody facilities in the U.S., asking for information on each youth assigned a bed in the facility on the last Wednesday in October.
Rates are calculated per 100,000 juveniles ages 10 through the upper age of each state's juvenile court jurisdiction.

(Continued)

Figure 5.2   (Continued)

Q: How long do juveniles stay in juvenile residential placement facilities?

A: Half of committed residents had been in placement longer than 120 days when the census was taken. Half of detained juveniles had been in placement fewer than 22 days.

Median days in placement since admission, by placement status, 1997-2013

Placement status	1997	1999	2001	2003	2006	2007	2010	2011	2013
All facilities	69	76	77	69	65	64	70	63	76
Committed	112	125	128	113	112	110	114	103	120
Detained	17	17	15	15	16	15	19	15	22
Private facilities	110	112	110	109	107	104	113	98	113
Committed	121	131	124	121	120	119	127	111	126
Detained	26	21	20	20	22	17	19	19	28
Public facilities	55	62	63	50	47	48	50	47	58
Committed	106	120	132	105	107	104	106	97	117
Detained	16	16	15	15	15	15	19	15	21

NOTES: The "median days in placement" statistic indicates that half the residents stayed fewer days and half stayed more days.

SOURCE: Adapted from National Council on Crime and Delinquency, *And Justice for Some*, Poe-Yamagata and Jones (2007).

question that for those states continuing to send more juveniles to adult prisons, the burden of this ill-conceived penal policy will be most heavily borne by young people of color.

BEYOND BLACK AND WHITE

The issue of racial disparity in the juvenile justice system is not limited to the experience of African American youths. Although the data presented in this chapter clearly demonstrate the extremely harsh treatment received by African American youngsters in the juvenile justice system, it is worth examining how youths from other minority communities are treated by the system. More research is needed on the plight of African American youngsters, but there is a virtual absence of data on youths who are Latinos, Asian and Pacific Islanders (API), and Native Americans. A recent literature review conducted by OJJDP found fewer than 15 articles over the last decade that cover the experiences of these significant ethnic communities and the justice system (Pope, Lovell, & Hsia, 2002). Statistical data that would help clarify key issues are almost nonexistent or are confusing at best.

One major problem is the way in which the federal government collects and reports statistical data on a range of ethnic communities. For example, the U.S. Bureau of the Census has only recently permitted people of mixed racial heritage to acknowledge that reality in the 10-year population counts. Further, the federal government imposes the arbitrary rule that Latinos (or *Hispanics* in their terminology) constitute an "ethnic category" whose members might be of any racial group. Why this distinction is reserved only for Hispanics and not other groups is not well explained. Moreover, federal data are extremely uneven with respect to more detailed breakdowns of subpopulations within the racial categories. This is becoming an increasing problem as the United States welcomes immigrants from a long list of other countries and these newer immigrants are lumped into statistical categories with long-standing immigrant populations. Most federal reports mix together groups that have little or nothing in common based on culture, citizenship status, life experiences, or language. For example, more than 50 nations have contributed to the U.S. population of Asian Americans, including ethnic groups that speak more than 100 distinct languages. The U.S. government adds to this broad category current and former residents of the Pacific Islands and Native Hawaiians. Despite the more accepted term of *Native Americans*, the U.S. Census continues to refer to *American Indians*, although the many ethnic groups within this category contain virtually no persons with historic connection to the Indian subcontinent. It has been more than five centuries since those hopelessly lost European explorers thought that they had sailed to India rather than the New World, yet our government continues to institutionalize their mistake. To make matters worse, the various Native American peoples are merged with a variety of indigenous groups who primarily come from Alaska and parts of Canada. These arbitrary and apparently meaningless statistical categories render virtually useless most of the data on these groups.

This data situation gets worse at the state and local levels. Every state has its own policies with respect to the racial and ethnic identifications used to compile statistics on crime and justice. There is virtually no training for law enforcement and other justice personnel on how to correctly identify an individual's racial or ethnic affiliation. Some agencies ask the defendant to self-identify, but others rely on guesses. Because the justice system employs far too few people of color, one cannot be too sanguine about the accuracy of these racial and ethnic identifications. Moreover, as indicated at the beginning of this chapter, the categories traditionally used in analyses of race and ethnicity are highly interconnected and subjective in their interpretations and practical impacts. Despite these enormous problems, it is worth examining some of the unique juvenile justice issues faced by Latino, Asian American, Pacific Island, and Native American youths.

¿Dónde Está la Justicia?

Despite the great difficulties in obtaining accurate data on Latino youths and the juvenile justice system, the Institute for Children, Families, and Youth at Michigan State University produced an extremely valuable summary of the issues (Villarruel & Walker, 2002). Similar to the findings for African American youngsters, Latino youth are overrepresented at virtually every stage of the justice system. They appear to receive harsher treatment than white youths, even if the alleged offenses are the same. As noted earlier, an NCCD study reported that Latino youths were more likely than their white counterparts to be sent to state juvenile correctional facilities, even when one controlled for the severity of the commitment offense and prior record (Poe-Yamagata & Jones, 2000). These disparities also existed for the lengths of stay that individuals from each ethnic group spent in state facilities. Data from Villarruel and Walker (2002) reveal that Latino youths are incarcerated in adult prisons and jails at 3 to 6 times the rate of whites in nine states, and 7 to 17 times the white rate in four states. An in-depth study in Los Angeles County, California, covering the period 1996 to 1998 showed that Latino youths were more than twice as likely to be arrested compared to white youths; they were more than twice as likely to be prosecuted as adults, and more than 7 times more likely to be sent to state prisons.

Solid information on Latino youths in the justice system requires a significant investment in both staff time and resources to improve the glaring gaps in data. There are also very important definitional and conceptual problems that must be solved. The federal government and many

CASE STUDY: RACE AND JJS

Shortly after passage of a new law that allowed youth who were as young as 14 years old to be tried as adults for homicide, Sean and Hector were both arrested in Long Beach, California. Sean, a white youth, took his father's gun and killed his mother in a dispute over cookies that he wanted. Hector, a Latino youth, was an accomplice to an armed robbery in which his codefendant shot and killed a convenience store clerk. Hector had no gun. He was basically in the wrong place at the wrong time. Neither youth had a history of prior violence. Both youngsters had public defenders. Sean's lawyer presented a very complete social history of this youth, but Hector's attorney did not even request a hearing. Hector's lawyer explained that the fear and hatred toward Latinos in Southern California would have made appeals for leniency futile.

Despite recommendations for each youth that they were suitable for rehabilitation in the juvenile correction system, Hector received a life sentence in the state prison system, but Sean was sent to the California Youth Authority. The sentencing judge in Sean's case listened intently to testimony that he suffered from posttraumatic stress, whereas the judge in Hector's case referred to him as a "cancer on society."

states do not have a uniform definition of who is Latino or Hispanic. Different agencies utilize different categories. Further, these categories are treated as overlapping with existing racial statistical designations—meaning that a Hispanic person may be of any race. Due to the fact that most Latinos are designated as white, the net result is an overestimation of the number of whites and an undercounting of minorities in confinement.

Currently, states and counties may or may not participate fully in federal data collection efforts, so that the effort aimed at achieving accurate categorization is extremely varied. Moreover, there are little or no funds made available either to train justice system personnel in collecting data on Latinos or to support supplementary data gathering on Latino populations. The absence of bilingual staff in the justice system poses another challenge to achieving accurate data collection. Traditionally, states with large Latino populations such as Arizona, California, Colorado, and Texas have made a greater attempt to assemble data on Latinos. But as this fastest-growing segment of the U.S. population continues to expand in many other states, especially in the Southeast and the Midwest, data collection has not kept pace. The consequence of these discrepancies in data collection is that comparisons among states on Latino youths are virtually meaningless.

Beyond the problems of data collection, the absence of bilingual staff in the justice system has profound and negative consequences for Latino youths and their families. It is estimated that as many as half of Latino youths possess limited proficiency in the English language and that this issue may even be greater for their parents (Villarruel & Walker, 2002). For many of these youths and their families, the failure to have adequate bilingual services means that legal documents go untranslated and unread. Youth and their families may not be fully aware of how to exercise their legal rights or to adequately participate in their own defense. Difficulties in communication among families and justice system personnel tend to exacerbate an already critical situation. Further, the failure to offer meaningful bilingual services may mean that various social and psychological assessment tools that are used by the court may be grossly inaccurate; identifying and diagnosing mental health issues is especially problematic. The limits on communication can also hamper the effectiveness of counseling, aftercare, and other treatment services, and fuel the perception that Latino youths are less amenable to home-based services, thus propelling them to out-of-home settings.

Language barriers are just the tip of the iceberg. Lack of understanding of cultural nuances within diverse Latino cultures may lead to harsher treatment of these youngsters. Villarruel and Walker (2002) point out that a downcast gaze in the presence of an authority figure is a sign of embarrassment in many Latino cultures but may be interpreted by the non-Latino juvenile justice person as a sign of disrespect or a lack of remorse. Moreover, given that the Latino community consists of a broad array of cultures, including Mexican, Puerto Rican, Cuban, Guatemalan, Salvadoran, and Nicaraguan, to name a few, there exists a wide range of more specific cultural values, family structures, socioeconomic statuses, and immigration histories. The ability or inability of the juvenile justice system to provide services that are "culturally competent" can help or hinder the system's ability to provide meaningful interventions and services. For example, those who have recently emigrated from countries with totalitarian regimes may feel an understandably strong hostility toward or fear of law enforcement officials. The ability to trust that one's family will be treated fairly by authority figures is key to the successful functioning of the juvenile justice system. When juvenile justice personnel appear to have no understanding or appreciation of various Latino cultures, it is difficult to build a foundation of trust.

Latino youths are also detained for both short and long periods by the Immigration and Naturalization Service (INS). Each year, almost 5,000 immigrant youths are confined in more than 90 INS facilities across the nation (Villarruel & Walker, 2002). While not all of these youngsters are Latino, the vast majority are from Mexico and other Central and South American nations. These youths have committed no crimes other than being in the United States without proper documentation. Many of these youngsters migrated with their families, but they are often separated from family members and held in different facilities. Detention may take a few days or many months. Parents may be reluctant to visit their detained children for fear that they themselves will be deported. These youths are given virtually no legal representation, nor are they told what is likely to happen to them. The INS often has few bilingual staff. Culturally competent programming is virtually nonexistent. Some INS facilities do not meet even minimum standards for housing children. Even worse, the INS record-keeping and computer systems are so inadequate that it is difficult to determine how long these youths remain in custody or what happens to them upon release.

Latino youths suffer disadvantages in the juvenile justice system due to questions about their immigration status. Youths who are citizens or who are in the United States via legal immigration are often presumed by law enforcement officials to be illegals. Latino youths are often stopped and searched by police using the sole rationale that they are potentially illegal immigrants. The assumption that a youth is breaking U.S. immigration laws can result in incarceration until parents or guardians are located. Also, a youth who is wanted by the INS may be housed in a juvenile detention facility until INS officials decide how they wish to handle that particular case. Such youngsters are not eligible for release to their parents or to any other alternatives to detention. These youths are in limbo—they are charged with no offense, but the legal process sits in suspension, sometimes for extended periods of time.

In the aftermath of the attack on the World Trade Center in New York City, the U.S. Justice Department encouraged local law enforcement agencies to assist in the work of identifying persons with questionable immigration statuses. This policy, if fully implemented, will result in even higher numbers of Latino youngsters being arrested and incarcerated. During the Obama administration, there was vigorous enforcement of U.S. immigration laws and a growing number of deportations. Youth who are swept up in these deportation raids are likely to be held in a variety of federal, state, and local facilities. Companies who supply private prisons are increasingly providing incarceration for those awaiting deportation, including young people and their caretakers (Krisberg, Marchionna, & Hartney, 2015).

Another serious burden faced by Latino youths is the popular stereotype that most are gang members. Even though there is scant evidence that Latino youths are more involved in gang activities than youths from other ethnic groups, media portrayals of the "Mexican bandit" have shaped public perceptions and law enforcement responses to Latino young people. Being labeled as a gang member increases one's likelihood of arrest and detention. Under certain state laws (such as California's Proposition 21), even the allegation that one is a gang member can result in a wide range of adverse consequences. Proposition 21 permitted a youth as young as 14 to be tried as an adult at the discretion of prosecutors. Offenses allegedly tied to gang activities permit automatic filing of juvenile cases in criminal courts. Many law enforcement agencies have implemented computerized gang intelligence systems. Every time a youth comes in contact with the police, officers may access this gang intelligence file. The problem with these systems is that there is no real legal standard for defining a person as a gang member. Having a relative or friend who is alleged to be part of a gang is enough to get a youth labeled. School officials have identified youths as gang members based on the clothes they wear, including the logos of most professional and college sports teams. Likewise, driving certain types of cars may lead school personnel or police to conclude that an individual is a gang member.

Even worse, once one is labeled as a gang member, it is nearly impossible to be deleted from these automated files. Youths and their parents are not informed that they are included in these files and have no means to expunge their names once in the database. Being labeled a gang member, for all too many Latino youths, is a lifetime disadvantage.

In juvenile correctional facilities, it is not uncommon for virtually all incarcerated youths to be treated as if they are active gang members. If a fight between two suspected Latino gang members breaks out, correctional officials may lock down all Latino youths in the facilities, whether they were part of the altercation or not. Housing assignments in prisons and juvenile facilities are sometimes dictated by staff perceptions of gang membership. A youth who is unaffiliated is faced with the difficult choice of joining a gang or being victimized by other gang members.

Hysteria about gangs has led to intensified police patrol in Latino neighborhoods and targeted police sweeps in these areas. Some cities, such as Los Angeles, have also enacted **gang injunctions** that prohibit presumed gang members from associating with one another—this can include family members. Latino youths are frequently harassed by police as probable gang members. The image of the Latino gang member is often merged with sinister racial stereotypes of the "illegal alien." In the aftermath of the September 11 tragedies, there have been several instances in which Latino youths were treated as if they were terrorists. There was a somewhat ambiguous case of a young Puerto Rican man in Chicago who had taken on an Islamic name and was allegedly found with some nuclear waste material. Although it now seems highly unlikely that this case was in any way related to jihadists, there was enhanced surveillance on Latino gangs in many cities. This is yet another example of how Latino youngsters are marginalized and feared by white Americans.

Asian Americans and Pacific Islanders: The Burden of Invisibility

Between 1977 and 1997, arrests of Asian Americans and Pacific Islanders (API) increased by 726% (Federal Bureau of Investigation, 1997). During this same time period, arrests of African Americans declined by 30%. The increase in API arrests far outstripped API increase in the general population (which virtually tripled from the 1980 Census to the 2000 Census). Although the FBI data do not break down the API arrests by specific ethnic groups within this population category, data presented below demonstrate that refugee populations from Southeast Asia were very involved with the justice system. Migrants from Southeast Asia contained high concentrations of war victims, refugees, and other dispossessed populations. Youths who are Samoan have very high rates of arrest and confinement in California (Asian/Pacific Islander Youth Violence Prevention Center, 2001). For example, Samoan young people had a higher arrest rate than African American youths in San Francisco. Southeast Asian youngsters and Samoan youths are much more likely to be placed in out-of-home settings than white youths. In the state of Hawaii, Native Hawaiian youngsters are disproportionately detained and imprisoned.

Vietnamese youths made up a highly disproportionate number of juvenile arrestees in both Alameda and San Francisco counties in California (Le, Arifuku, Louie, & Krisberg, 2001; Le, Arifuku, Louie, Krisberg, & Tang, 2001). Statewide, the number of Asian youths committed for the first time to the California Youth Authority (CYA) rose from 4% to 7.4% of the total new commitments during the 1990s (State of California, Department of the Youth Authority, n.d.). A survey of the February 2002 residents of CYA found that 25% of the API youths in custody were committed for homicide, compared to 6% for whites, approximately 9% for Hispanics, 12% for Native Americans, and 8% for African Americans. Thai, Laotian, and Cambodian youths from California's Central Valley were overrepresented in the CYA population. Although API groups have been left out of many studies of self-reported delinquency, the limited available data suggest that Southeast Asian youths self-report higher rates of antisocial behavior than non-API youths (Asian/Pacific Islander Youth Violence Prevention Center, 2001). A recent report suggests that young women from Southeast Asian communities have the highest rates of teenage pregnancy compared to girls from other ethnic groups (Asian/Pacific Islander Youth Violence Prevention Center, 2010).

These facts are striking in contrast to the dominant perception of API youths as members of the "model minority." A very pervasive view of the API population is that they are a racial minority group that has overcome its disadvantaged status and achieved a very high degree of educational success, financial independence, and social acceptance. The myth of the model minority is that APIs prove that strong family values, a strong work ethic, and a commitment to educational attainment can overcome generations of racism and legal discrimination against Asian Americans (Lee & Croninger, 1996; Sue & Kitano, 1973). The presumed success of the API population is offered as evidence of the powerful existence of the American meritocracy that transcends racial or class differences. Other ethnic groups are advised, by implication, to follow the lead of the API population to find the keys to advancement in American society. API achievement has been used to attack affirmative action programs, social welfare policies, and the decline in family values among other ethnic groups. It is certainly true that some segments of the API population are doing better after years of suffering under racial discrimination and legally imposed barriers; however, there are many segments of the API community that are not succeeding at all (Takagi, 1989). For example, the 1990 Census reported that Laotian and Hmong

Americans had rates of poverty that were more than 6 times those of the general U.S. population. Cambodian Americans had a poverty rate of 47%, and Vietnamese Americans had a poverty rate of 34% (President's Advisory Commission on Asian Americans and Pacific Islanders, 2001). Whereas 38% of all Asian Americans held college degrees in 1990, this was true for only 6% of Cambodian Americans, 7% for Laotian Americans, 3% of Hmongs, and 17% of Vietnamese Americans (Sok, 2001).

The myth of the model minority means that very needy segments of the API community are ignored by government agencies and social services agencies (Lee & Zahn, 1998). Moreover, the API population exhibits very low rates of voting and consequently has very few elected representatives at any level of government (Espiritu, 1992). With few exceptions, API populations live in ethnically segregated communities along with other disenfranchised ethnic groups (Ong & Miller, 2002). This further contributes to the political invisibility of the API population.

API youths are confronted with many of the problems discussed above in connection with Latinos. Despite the growing numbers of API youths entering the justice system, there have been few, if any, accommodations to the special needs of the API population. There are very few police, judges, prosecutors, defense lawyers, or correctional officials from the API community and even fewer justice officials. Language barriers are enormous for groups that together speak over 100 different languages and dialects. Interpreters are virtually nonexistent. Even in jurisdictions such as San Francisco, in which APIs constitute almost a majority of the youth population, there has been little attention paid to developing culturally competent programming.

API youths also suffer from a range of racist stereotypes in their encounters with justice system officials. Contemporary law enforcement officials are fond of promulgating the images of the evil Chinese gangs that originate in Hong Kong, Taiwan, or China and victimize American citizens. These images are as old as those presented to the gullible public in the late 19th and early 20th centuries by the Hearst and McClatchy newspaper chains that showed cartoons of sinister Asian men luring attractive white women into opium dens, forcing them into lives of degenerate white slavery. These racist images set the stage for a wide range of anti-Asian legislation and exclusionary immigration policies. The mass media has trumped-up the Asian American criminal as one of its most frightening images.

Immigration issues also plague youths from API communities. They face many of the same kinds of injustice and insensitivity encountered by Latino youngsters. The image of the Asian youth gang leads police to arbitrarily stop many API youths. Laws designed to crack down on gangs are disproportionately applied in API communities.

NATIVE AMERICAN YOUTHS: OUTSIDERS IN THEIR OWN LAND

Ironically, there are less data available on the involvement of Native American youths in the juvenile justice system than for any other major ethnic group. Although the FBI reports that there were 20,295 "American Indian or Alaskan Native" juveniles who were arrested in 1999, it is impossible to determine the accuracy of this number. Native Americans reside both inside and outside **tribal territories.** Some of the tribal areas are regarded as sovereign nations; others are not. The bulk of the juvenile arrests reported by the FBI are from urban areas, suggesting an underreporting of arrests occurring in tribal areas (Maguire & Pastore, 2000).

Melton (1998) points out that reliable statistics on Native American crime are missing for several reasons. First, data collection methods differ from tribe to tribe, and there are no uniform methods governing these data. Second, many crimes that occur in tribal territories are not even reported to the FBI. Other federal data on Native American crime are collected by the Bureau of Indian Affairs (BIA) or U.S. Attorneys on cases under investigation, but these data are very incomplete. Law enforcement in tribal areas is provided by a mixture of local, state, and federal agencies. According to Melton (1998), about 60% of 304 Native American reservations operate their own tribal police with financial support from the federal government. The BIA of the U.S. Department of the Interior has only about 400 sworn law enforcement officers that are assigned to about 40 locales, primarily in the western states. The BIA also has criminal investigators that support the work of these police officers. As mentioned earlier, there are circumstances in which U.S. Attorneys, the FBI, and other federal law enforcement agencies are involved in policing Native American territories. Observers have raised concerns about the poor level of coordination among these multiple law enforcement agencies, which contributes to the absence of reliable data about crime and the operation of the justice system in tribal areas. There are also concerns that services to Native American victims are inadequate, prosecutions are often hindered by confusing jurisdictions, and law enforcement training is uneven (Melton, 1998).

Adult and juvenile correctional facilities on tribal territories are operated by the BIA. In 1995, there were a total of four juvenile facilities operated by the BIA. There are also nearly 40 adult facilities that have some capacity to hold juveniles. In all, there many be as many as 339 beds to hold Native American juvenile offenders (Melton, 1998). Besides the correctional beds operated by the BIA, there are a number of contracted facilities used to house minors. There are very little reliable data on the programs and conditions of confinement in these contracted facilities. Some Native American youths who are convicted of federal offenses are transferred to the custody of the Federal Bureau of Prisons, which in turn contracts with state and private agencies to hold these youngsters. One consequence of this arrangement is that Native American youngsters are often confined in facilities that are hundreds of miles from their home communities. For example, Navaho youths from Arizona are often sent to the facilities of the CYA. It is rare that these contracted programs have either staff or programs that are culturally appropriate for Native American youngsters.

Native American families and youths confront many serious problems related to youth crime. Substance abuse, especially alcoholism, is a major problem. Child maltreatment is also a major problem in tribal areas (Hammond & Yung, 1993). There are increasing reports of the growth of gang membership among Native American youths (Coalition for Juvenile Justice, 2000; Melton, 1998). Community resources that are available to respond to these challenges are extremely underfunded and in some instances nonexistent. Community-based options, home-based services (in lieu of confinement), or out-of-home placements are badly needed. Not surprisingly, this situation leads to disproportionate incarceration among Native Americans, both in tribal areas and in the outside community. Because many Native Americans live in rural communities, the excessive confinement of these youngsters is further propelled by the general lack of alternatives to incarceration in rural areas, as well as the punitive philosophies that tend to dominate rural justice systems (Parry, 1996).

The task of bringing justice to Native American youths is enormous. There is a paucity of solid research to guide program development. There are few well-tested models of working effectively with Native American youngsters and their families. We need to travel the long journey to educate justice system officials about culturally appropriate interventions. The severe balkanization of law enforcement services and correctional programs must be reduced.

As with API and Latino youths, we must focus positive public attention on these children and families. The American legacy of violence against and exploitation of its native peoples is an issue of enduring national shame. The continuing plight of Native American youths should not be allowed to perpetuate that tragic history.

IN SEARCH OF ANSWERS

In our society, race and ethnic differences affect virtually all aspects of life. One's racial or ethnic identity exerts a major influence on where one lives, the quality of health care one receives, where one's family attends religious services, the friends that one associates with, the neighborhoods that one is likely to live in, the range of educational and vocational opportunities available, and the quality of justice one can expect. Indeed, race and ethnicity affect a person's life expectancy and where bodily remains will be placed after death. It would be odd if the juvenile justice system were *not* profoundly impacted by race and ethnicity.

Though there is little doubt that American society has made some significant strides to eliminate legally sanctioned segregation and discrimination, the remaining disparities are nevertheless extremely important. America's long and tragic history of racism is still with us. Like a virus, we are all infected by racist concepts and fears. The virus may lie dormant for a time, but it is nearly impossible to escape its pernicious influence. The benefits that accrue without effort to all white Americans (sometimes referred to as *white privilege*) is a seductive reason to deny that racism still exists or that one has any specific responsibility in the nation's legacy of racial oppression. Some of the racial disparities discussed in this chapter are the result of prejudiced individuals who consciously set out to harm people of color. If these individuals were the major source of the problem, the remedies would be rather straightforward—screen out biased criminal justice personnel and punish those who use race or ethnicity as inappropriate criteria in the exercise of their legal responsibilities. Unfortunately, the issue is far more complex, and the unconscious influence of racism is far more pervasive and difficult to purge from the operations of the justice system. One example of the effects of unconscious race bias is offered by the research of Bridges and Steen (1998). They systematically reviewed the court reports produced by probation officers in a northwestern urban juvenile court. The findings were startling. The probation officers were much more likely to attribute the delinquency of white youths to environmental factors, of which the youths were victims. For black youths, the probation officers concluded that the misconduct was due to fundamental character flaws. It was typical to read that black youths "showed no remorse" or were "cold-blooded" offenders. The white youths were often portrayed as tragic figures of family maltreatment, mental health problems, or bad companions. Bridges and Steen found that these probation officer analyses differed even when the circumstances of the offenses were almost identical. Similarly, the probation officers were more likely to recommend incarceration as the appropriate punishment for African American youngsters. They were more willing to consider home-based treatment approaches for white youths.

The contrasting stories of two youths from Southern California further illuminate the forces that influence disparate treatment of minority youths by the justice system. In 1995, the California legislature enacted a new law that permitted youths as young as 14 to be tried in adult courts for murder. (It should be noted that since that time, California voters have voted for even more expansive procedures to prosecute very young children in criminal courts.)

The first two youths who were tried under the 1995 statute were both from Orange County, California. While the names of the youths have been changed to protect their confidentiality, the facts of the case illustrate how racial and ethnic bias can corrupt the justice process (Krisberg, 2003). The first youngster, whom we will call Sean, is a white youth. Sean took his father's gun and killed his mother in a dispute over cookies that Sean wanted to eat. The second youth, whom we will refer to as Hector, was Latino. Hector was an accomplice to an armed robbery in which his codefendant murdered a convenience store clerk. Hector had no gun and fired no shots. He was simply at the wrong place at the wrong time. Both youngsters were 14 years old at the time of these events.

Under the 1995 law, both youths were eligible to be tried as adults, but the outcomes of these cases were quite different. Hector was tried in a criminal court and is serving a prison sentence of 25 years to life. Sean was kept in the juvenile court, was sent to the CYA, and was eligible for parole in a few years. Both Hector and Sean were initially sent to a Youth Authority Reception Center for a diagnostic assessment. Sean had no prior criminal record, whereas Hector had a few prior arrests for minor property crimes. Neither youth had an official history of past violent behavior. Correctional professionals from the Youth Authority concluded that both young people were amenable to the treatment programs of the juvenile justice system.

Sean became the focus of national media coverage. His father appeared on the ABC show *Nightline* pleading for leniency for his 14-year-old son. Hector's case achieved minimal media attention, with only a very brief report in the Orange County version of the *Los Angeles Times*.

During his hearing, Sean was portrayed by his counsel as a tragic and sympathetic figure in need of mental health treatment services. In court, Sean, a tall and muscular high school football player, acted contrite and despondent. Hector, who was very slight of build, swaggered and grinned, showing the adolescent bravado that he had learned on the streets of Long Beach, California. The judge in Hector's case referred to him as "a cancerous growth on society." Neither boy said he could remember the details of his crime. Hector's lack of memory was interpreted by the prosecution as "a lack of remorse." Sean's claim of not recalling the murder was explained by a court-appointed psychiatrist as "post traumatic amnesia" (Krisberg, 2003).

Neither family had much money, and both boys were represented by public counsel. In Sean's case, he was defended by a very able public defender that put on a "virtual capital defense" to allow the court to see him as a sympathetic figure. Sean's public defender asked the court to pay for a psychiatrist who examined the troubled 14-year-old and made a convincing case for leniency. The court heard from a broad range of adults who knew Sean his whole life. By contrast, Hector was given an assigned counsel, a private lawyer who periodically takes on cases of indigent clients. In general, the assigned counsels are somewhat less familiar with the local juvenile court culture. Unlike Sean's adroit public defender, Hector's lawyer did not request a court-appointed psychiatric exam and did not put on a lengthy case. In fact, there was no formal court hearing for Hector; his plea for leniency amounted to the filing of papers.

Sean clearly benefited from skillful legal representation as well as community sentiments favoring compassion and understanding. Hector was not as fortunate. As one observer reported, "There is a sea of Hectors in Southern California, and we can hardly even distinguish them as individuals" (Krisberg, 2003).

Blatant racism would be an easy explanation for the disparate outcomes between Hector and Sean. After all, the white youth got the benefit of the doubt and the Latino youth was harshly punished—even though the offense committed by Sean seems much more frightening than that of the unarmed codefendant, Hector. The racially charged atmosphere in Southern California, especially the bias against Latino citizens, surely contributed to the end result. Clearly, there were two standards of justice that were applied in these two cases. Yet it is unlikely that the criminal justice officials set out to conspire against Hector. More realistically, unconscious race and class prejudices came into play. No less pernicious, racism that exists below the cognitive level can exert a very powerful influence on human decision making (Lawrence, 1987). One might raise the broader issue that unarticulated racial biases were crucial to convincing legislators that children as young as 14 should be tried in adult courts (Krisberg, 2003).

The cases of Hector and Sean provide humanistic texture to the social science research of Bridges and Steen (1998). These data also point us in the direction of possible remedies to the unfair treatment of minority youths at every level of the juvenile justice process (Krisberg et al.,

1987; Poe-Yamagata et al., 2007). One such solution, which seeks to make decision makers more aware of their unconscious biases, involves training in racial and ethnic sensitivity. Diversity training has become increasingly important in criminal justice training curricula. Though we have little completed empirical research on the effectiveness of these training investments, there is scant evidence that actual decision-making practices have changed in very many locales. Typically, justice system workers deny that they are using inappropriate decision-making criteria; they put forth a range of explanations and justifications for unfair outcomes. Most prominent is the erroneous assertion that solving racial bias in the justice system will require the imposition of a quota system for who gets arrested and detained. Diversity trainers confront predominantly white justice system employees who feel defensive about claims that the system is racially slanted. Moreover, diversity training is often subtly presented as something to be done for public relations reasons. Frontline workers often do not get a clear message that top management is totally committed to changing policies and practices.

More promising results have been reported in jurisdictions that have an explicit goal to reduce minority penetration into the juvenile justice system, especially in those places that have moved to more objective decision-making tools. For example, two sites supported by the Annie E. Casey Foundation, Santa Cruz, California, and Multnomah, Oregon, have achieved impressive results in lessening racial and ethnic disparities in their detention centers (Hoytt, Schiraldi, Smith, & Ziedenberg, 2001). The Building Blocks for Youth project is now working actively with a number of jurisdictions that are willing to make a commitment to reducing racial disparity and that are willing to examine every level of decision making throughout their systems. NCCD's research has consistently demonstrated that research-guided decision systems can reduce the racial disparities that seem endemic to clinical or subjective systems in both juvenile justice and child welfare (Baird, Ereth, & Wagner, 1999).

SUMMARY

In 1992, the federal JJDPA was amended to require all states wishing to receive funds from OJJDP to examine the extent of DMC in their juvenile justice systems. In addition to conducting the studies, states were required to make good-faith efforts to reduce DMC. Many of these studies have documented the disproportionate presence of children of color at every stage in the juvenile justice process. In particular, African American, Latino, and Native American youngsters are much more likely than white youths to be detained, to be placed out of the home, and to be sent to prisons.

Research that explains why these disparities occur is less well defined. The available data on race and ethnicity and juvenile justice are very uneven. It appears that many factors contribute to an "accumulated disadvantage" that pulls children of color into the deep end of the juvenile justice system. There are previously ignored issues that adversely affect the handling of Latino, Native American, Asian American, and Pacific Islander youngsters. These include inadequate language and cultural resources in the juvenile justice system, excessive focus on gangs, racial biases that are embedded in juvenile justice decision making, and detrimental policies of the INS and the BIA. There are several promising approaches to reduce DMC, and some jurisdictions have made some significant progress in this area.

The W. Heywood Burns Institute has many training and technical assistance projects to reduce the extent of racial disparity in the justice system. In particular, this institute has found that juvenile justice workers falsely believe that the law requires them to resort to the most severe sanctions, even though the reality is that most state laws provide a very wide range of discretion throughout the juvenile justice process. The Burns Institute is a great resource on data, research, and innovative policies that can reduce DMC (Burns Institute, 2015).

Another reform direction suggested by the research and case examples is to increase the quality and extent of legal representation that is available to all youths. Often, this is an issue of money—who can pay for adequate legal representation or which jurisdictions provide the best quality of legal defense of indigent clients. Research and demonstration projects need to be launched to test whether improved legal services can help reduce the racial biases of the juvenile justice process.

Because the decision to transfer youths to the criminal court seems to have an especially significant impact on minority youths in the justice system, it makes sense to oppose the expansion of transfer practices and even attempt to reverse some of the very bad laws that have been enacted during the past decade. Franklin Zimring (in press) has argued that focusing on reducing the growing trend to try juveniles as if they were adults is a "harm reduction" strategy that will positively impact minority youngsters without explicitly challenging the racial biases in decision making.

Finally, it appears that early interactions between minority youngsters and the police or detention intake workers are crucial to the ultimate disparate outcomes. It is very important to create a range of new community-based options that will be available in the communities in which most children of color reside. Diversionary programs have too often been instituted only in predominantly white communities. Specific attention should be paid to developing inner-city program options to keep nonviolent offenders out of the justice process whenever possible.

Research pointing to proven alternatives is woefully inadequate. We need to know more, but we also need to *do* more. Large-scale demonstrations of how to reduce racial disparity in the juvenile justice system have not been undertaken. Simple justice demands a course of action different from the current one. Indeed, if the reality of the justice system for children of color is that "justice means just us," the very legitimacy of the legal process is compromised. There are few issues more central to the reform of juvenile justice than rectifying the racial biases that seem to infect every aspect of the process.

REVIEW QUESTIONS

1. What is the best way to measure the extent of DMC, and what stages of the juvenile justice process are most important in generating disproportionate treatment of children of color?

2. What is "cumulative disadvantage," and how does it increase DMC?

3. What are some of the key conceptual and data problems that limit our ability to really understand the causes of DMC?

4. What are the particular challenges in the justice system that are confronted by Latino, Native American, and Asian and Pacific Islander youngsters?

5. What approaches have proven successful in reducing the extent of DMC?

INTERNET RESOURCES

The National Council on Crime and Delinquency has produced many studies and reports on the impact of race and ethnicity on the justice system.
http://www.nccdglobal.org

The W. Haywood Burns Institute is a leading resource on research, policies, and technical assistance on reducing racial disparities in the juvenile justice system.
http://www.burnsinstitute.org

John Jay College of Criminal Justice has a strong research and educational focus on race and the justice system.
http://www.jjay.cuny.edu/

Chapter 6

YOUNG WOMEN AND THE JUVENILE JUSTICE SYSTEM

Of the approximately 1.6 million arrests of persons younger than age 18 in 2010, about 29% involved young women. Whereas young women were 18% of those arrested for violent crimes and 38% of serious property crimes, they were 82% of juveniles arrested for prostitution and commercialized vice, and 58% of those arrested as runaways. Girls accounted for about 16% of juveniles who were arrested for drugs and 11% of those arrested for weapons offenses (Sickmund & Puzzanchera, 2014).

In 2010, young women comprised almost half of the youth population ages 10 to 17. Data from self-report surveys suggest that young men and women are about equally likely to engage in sexual behavior, underage drinking, and the use of illicit drugs (Snyder & Sickmund, 1999).

Since the 1980s, the arrest rate of young women has been increasing at a rapid rate compared to that of their male counterparts, and they are being drawn into the juvenile justice system at younger ages (Sickmund & Puzzanchera, 2014). In 1980, girls were almost 17% of arrested juveniles, but that proportion increased to 29% in 2010. For the violent crimes, the proportion of females of juvenile arrests rose from 10% to 18%, especially for aggravated and simple assaults, and for very serious property offenses, the proportion of girls more than doubled. By contrast, the share of juvenile drug arrests for drug offenses was almost unchanged from 1980 to 2010. As noted earlier, the juvenile arrest rate has declined, but the drop for males was larger than for young women. Researchers are not certain as to the reasons for the differences in arrest trends. There is a view that changed law enforcement practices may be responsible for the observed trends. For example, there is speculation that police are now charging young women with assault charges in lieu of prior practices of treating family disputes as the juvenile status offense of "incorrigibility" (Zahn, Hawkins, Chiancone, & Witworth, 2008). The Office of Juvenile Justice and Delinquency Prevention (OJJDP) Girls Study Group proposed a series of research studies that might explain these trends in arrests of young women (Zahn et al., 2008).

The caseload of juvenile court cases involving females increased 69% between 1985 and 2010, compared with a 5% increase for males (Sickmund & Puzzanchera, 2014). This increase of girls in juvenile court caseloads was especially pronounced for crimes against persons and public order offenses. More recently, from 2001 to 2010, juvenile court caseloads have declined, but the drop in males in juvenile court was larger than for young females.

This significant increase in the number of young women being arrested and brought into the juvenile justice system has fueled considerable interest in the underlying forces behind this trend. At least one controversial book has attributed this relative rise in arrests among young (and older women) as a by-product of the women's liberation movement (Adler, 1975). However, a number of studies designed to test this thesis have not provided much empirical support. Other popular and media explanations of the increase in crime among young women include the effects of the War on Drugs or the trend toward the earlier physical maturation of young women. Some have asserted that the trends in official data are reflective of law enforcement practices in which behaviors that were traditionally labeled as juvenile status offenses are now treated as criminal offenses—a practice known as "bootstrapping" (Chesney-Lind, 1997). It is not uncommon for girls who are placed on probation as status offenders to be subsequently detained for the offense of "violating a valid court order." Thus, a young woman who initially enters the juvenile justice system for truancy may have her legal problems exacerbated if she violates a judge's order to attend school. Youth advocates claim that bootstrapping is a way in which juvenile courts can get around the legislative barriers to incarcerating status offenders. The severity of delinquent acts by girls is also exaggerated by relabeling family disputes as criminal events. For example, a study of girls who were referred to the Maryland juvenile justice system found that almost all of the young women referred for "assaults" involved family-centered arguments (Mayer, 1994). Young women have been inadequately represented in most large-scale and well-designed studies of delinquency. Consequently, there is very little sound research evidence that could be utilized to adequately describe or analyze changing patterns of female delinquency (Hoyt & Scherer, 1998; Zahn et al., 2008).

As noted earlier, girls are underrepresented in arrests of juveniles for violent offenses, but they are overrepresented in arrests for juvenile status crimes such as running away, incorrigibility, and curfew violations (Snyder & Sickmund, 1999). Several commentators have suggested that juvenile crime enforcement patterns appear to be protecting society from the criminal behavior of young males, but the same juvenile justice system seems focused on protecting young women from themselves (Bishop & Frazier, 1992; Chesney-Lind & Shelden, 1992; Mann, 1984; Weiss, Nicholson, & Cretella, 1996). Female delinquency has been traditionally interpreted by justice system agencies as wrapped up with concerns about sexuality and moral depravity (Hoyt & Scherer, 1998).

Data on the processing of girls by the juvenile justice system reveal a pattern and practice of handling delinquency cases via informal and diversionary programs (Poe-Yamagata & Butts, 1996). However, there has been speculation that some of these diversionary options may involve equally restrictive sanctions, albeit occurring in the welfare, mental health, or special education systems. Girls are far more likely than boys to be placed in privately operated group homes, shelters, and other residential programs. On the other hand, it is clear that girls are often detained and placed out of home for minor offenses that would not result in the same disposition for boys (American Bar Association & National Bar Association, 2001; Weiss et al., 1996).

Partly because boys are more numerous in the juvenile justice system, especially in locked correctional programs, there are few, if any, specialized programs for young women. A popular comment is that current girls' programs involve little more than taking a boys' program and "painting it pink." Such programs are often guided by gross stereotypes about the adolescent development of young women (e.g., girls like tea parties, young women are inherently deceitful, the principal focus of girls' programs should be helping them find an acceptable mate, and girls need to learn to control their sexual urges). Moreover, girls' programs are often given less budgetary resources, and young women are given second-class access to educational, vocational, and recreational resources in the juvenile corrections system. For example, a recent investigation of a co-correctional facility in California revealed that the young women had to wait for males to complete their recreational activities before they were allowed access to any of these facilities. This practice often resulted in the young women getting less physical exercise time than required by state policies. Other studies have pointed to gross deficiencies in the allocation of medical and other treatment resources to incarcerated young women. Programming for young women, especially in detention centers and other confinement settings, is severely underdeveloped. These same studies have uncovered patterns of verbal, physical, and sexual abuse of young women in the juvenile corrections system (Acoca, 1998; Acoca & Dedel, 1998). Revelations of the widespread sexual abuse of girls in juvenile corrections facilities led to closure of the Florida Institute for Girls and led to indictments of staff at the Marion County Indiana Girls' School (Krisberg, 2016). A national survey of youth in custody conducted in 2008 to 2009 related that 12% of girls reported being sexually victimized by staff or by other youth in the prior 12 months. About half of these incidents involved threats and intimidation. Almost half of these young women also reported sexual victimization in correctional facilities prior to the last 12 months (Beck, Harrison, & Guerino, 2010). The proportion of girls who reported sexual victimization was similar to that of incarcerated boys.

It has also been noted that there is lack of prevention and early intervention services for at-risk girls (Weiss et al., 1996). The low numbers of young women relative to young men who enter the juvenile justice system have been used to excuse the lack of services. There are growing voices to greatly increase the quality and availability of gender-specific services for girls who enter the juvenile justice system (Pasko, Okamoto, & Chesney-Lind, 2010).

GENDER-SPECIFIC JUVENILE JUSTICE SERVICES

Despite this bleak picture of services for at-risk young women, there is growing awareness and concern about the need to do better. For example, a number of states have convened task forces on gender equity in the court system, and these groups have focused more attention on the disparate treatment of women, both juveniles and adults, in the judicial system. Professional groups such as the American Bar Association and the National Bar Association (2001) have stressed equity issues and attempted to rally advocacy resources for young women who are caught up in the juvenile justice system. Others such as the National Council on Crime and Delinquency (NCCD), the American Correctional Association, and Girls Incorporated have issued detailed reports of the plight of incarcerated young women, and they have offered specific recommendations for action (OJJDP, 2015; Weiss et al., 1996; Zahn et al., 2008).

The growing professional consensus is that a priority must be placed on (a) ensuring equitable treatment of young women in the juvenile justice system and (b) establishing high-quality gender-specific services. The goal of equity requires that juvenile justice officials and legal advocates for young women purge sexual and gender stereotypes from the operation of the justice system. This includes ending the dubious practice of incarcerating young women for their "self-protection." Jurisdictions must carefully examine the disproportionate use of status offender laws to pull girls into detention and out-of-home placements. The practice of using court orders or bench warrants to bootstrap status offenders into allegedly more serious incidents of misconduct must be eliminated.

Using more objective detention and placement screening tools should reduce the numbers of inappropriately confined young women. These instruments highlight the fact that most girls who are detained represent a very low risk to public safety in terms of violence or further reoffending. The available research indicates that many young women in the justice system possess very high profiles for treatment needs. While the juvenile court should attempt to provide the necessary treatment interventions, there must be a higher priority placed on home-based or community-located services. Most important, the services that are provided must be developmentally appropriate for young women—no more boys' programs that are painted pink.

Several exploratory studies have pointed to crucial service needs for young women. For example, Acoca and her associates have documented that court-involved young women have been exposed to tremendous amounts of violence in their home environments and at very young ages (Acoca, 2000; Acoca & Dedel, 1998). A profile of young women in a gender-specific probation program in Alameda County, California, produced parallel results. Boys who engage in violent behavior also have been subjected to high levels of violence in the home, either as victims of abuse or as witnesses to it (Widom, 1989). For young men, it appears that the "cycle of violence" leads them into more aggressive behavior with peers as well as violence against intimate partners. For girls, heavy exposure to violence seems to lead to self-abusive behavior, maltreatment of their children, depression, mental illness, drug use, and suicide (Baerger, Lyons, Quigley, & Griffin, 2001; Freitag & Wordes, 2001; Hoyt & Scherer, 1998; Obeidallah & Earls, 1999). Studies are also beginning to document very high rates of sexual violence and abuse among girls in the juvenile justice system (Acoca, 1998; NCCD, 2001).

Researchers have pointed to etiologic factors such as early educational failure, family disruption, drug abuse, and gang involvement as decisive to propelling young women toward chronic and serious lawbreaking. Although these factors have also been identified as causal factors in male delinquency, there are crucial differences in the way in which these forces differentially impact males and females (Acoca, 1999; Brown & Gilligan, 1992; Dryfoos, 1990; Hoyt & Scherer, 1998; Moore & Hagedorn, 2001). These findings are interesting; however, it is vital to point out that the scientific literature on delinquency among young women is woefully incomplete. For instance, although there may well be risk factors that are common in delinquent pathways to both girls and boys, we have few clues to more subtle ways in which these factors interact over the course of adolescent development.

Another major knowledge gap involves protective factors that might insulate young women from adverse social and environmental forces or those steps that we could take to build resiliency for young women who have already been victimized and harmed (Greene, Peters, & Associates, 1998; Poe-Yamagata, Wordes, & Krisberg, 2002). There is some speculation that there are many areas in which protective factors could be found. For example, it has been assumed that positive female role models would be especially useful to young women who are struggling with losses of self-esteem, those who lack caring and concerned adults in their lives, and those who are looking for affirmative examples of successful transitions from adolescence to adulthood. Another area that has been identified involves teaching appropriate and effective relationship skills (Loeber & Hay, 1997). Related to relationship-building skills is knowledge about how to identify destructive and potentially dangerous personal relationships.

Other national organizations have pointed to the loss of self-esteem experienced by young women during the teen years (American Association of University Women, 1991). It is argued that boosting self-esteem and counteracting negative self-images can insulate girls from some risk factors. Related to this are efforts to promote individualism and a sense of personal boundaries among at-risk young women. Also important is the development of a sense of hope for a positive future (Weiss et al., 1996). There has been increased emphasis on the productive role of sports and other physical challenges in helping at-risk teens through difficult life transitions.

While many of these ideas appear to make sense on the surface, there is virtually no research supporting the value of these efforts as protective factors. It is easy to see that some adolescent boys could benefit from the same sorts of preventive services. There are virtually no proven program models that could demonstrate how best to deliver these protective factors. Further, researchers have not been given adequate support to determine if these positive protective factors should be delivered in culturally relevant ways. This latter point is especially important given that the vast majority of girls entering juvenile correctional programs are young women of color (Morris, 2002).

WHAT WORKS FOR AT-RISK YOUNG WOMEN

Reviews of tested juvenile justice programs have rarely, if ever, covered programs for young women (Krisberg, Currie, Onek, & Weibush, 1995; Lipsey & Wilson, 1998). Similar to the literature on prevention, some writers have speculated about what sorts of services and interventions might help at-risk young women make the transition to law-abiding conduct. These programs include pregnancy and parenting classes, mentoring and tutoring, opportunities for self-expression involving music and the arts, and wilderness challenges. There is virtually no research that backs up these ideas. At this writing, I know of only one program evaluation that incorporated an experimental design for a gender-specific program in Alameda County, California (NCCD, 2001). It is worth reviewing the results of that research.

With funding made available from the state legislature, the Alameda County Probation Department launched a multiyear effort to develop a model approach to gender-specific services for girls on probation. The RYSE program (Reaffirming Young Sisters' Excellence) was named after a contest among its early participants. The central idea behind RYSE was to offer a complete array of gender-specific services to girls on probation. It was designed as a model of home-based services, not as a residential program. It was hoped that RYSE would become a meaningful sanctioning option for the juvenile court and might avoid out-of-home placements for some young women. All girls placed on probation were eligible for RYSE. A random selection process monitored by NCCD then determined if the girls were assigned to RYSE or regular probation caseloads. Two thirds of the eligible girls went into RYSE and the remaining were assigned in the usual manner. Probation caseloads in Alameda County generally exceeded 75 clients per officer. Because males predominated in the probation population, there were rarely more than a few girls in any one officer's ongoing cases. It was planned that girls would stay in RYSE for 12 months and then be transitioned back to conventional probation supervision. With the leadership of the presiding juvenile court judge and the chief probation officer, it was decided that RYSE should be subject to a rigorous scientific testing. Generous funding from the state and county made the labor-intensive research possible over a 4-year period.

Like most juvenile justice programs, RYSE had multiple components, making it quite difficult to pinpoint the exact mix of services that were the most effective. RYSE clients were assigned to very small caseloads (usually less than 25 per officer) of specially trained probation staff. The girls saw their probation officers on a weekly basis. In addition, the RYSE clients attended a 12-week educational course called Sister Friends that was taught by community volunteers and the probation officers. The curriculum included a broad range of subjects including self-esteem, women's health issues, social skills, educational and vocational topics, reproductive health issues,

> ## CASE STUDY: YOUNG WOMEN AND THE JUVENILE JUSTICE SYSTEM
>
> Amalia may have been sexually abused as child, but she refuses to talk about it. Her mother died when she was 10 and her father left the home when she was very young. She has barely attended school and tries to care for her young siblings who sometimes live with her in a shelter. She likes sports and art.
>
> Amalia regularly uses alcohol and other drugs. She has spent time in several foster homes and is now pregnant. She experiences severe bouts of depression and has been diagnosed as bipolar. Her body reveals several scars from prior efforts to harm herself. She has no marketable job skills and no consistent work history.
>
> She goes to church every week and plans on raising her child herself. She admits to having a serious anger management problem. She stays in touch with an elderly grandmother with whom she lived for a while. She has a boyfriend who buys her clothes and tries to get her to engage in prostitution.

managing family conflicts, leadership roles for young women, and the arts. Sister Friends included field trips so that the young women could test out their new skills in positive settings that emphasized the achievement of women from their communities.

Each probation officer had access to funds to purchase appropriate treatment services, especially drug and alcohol treatment, and mental health services for their clients. There were also dollars available for "concrete services." The probation staff could access these funds for direct emergency financial assistance for their clients (e.g., paying a month's rent, covering overdue utility or phone bills, or purchasing needed school supplies or clothing). The idea of concrete services had been used by the probation department in its Family Preservation Unit, which was working to keep young people out of costly foster care placements. Utilizing the vouchers for services or the concrete service dollars required that the probation officer develop an overall **case management** plan that explained how these funds would support the youth's successful adjustment. In addition, the probation department contracted with local mental health professionals to assist the officers in dealing with especially difficult client and family situations. County health officials provided medical exams for those girls who had not received adequate medical care during the previous year.

RYSE served more than 450 young women over 3 years. As could be expected, the clients of RYSE exhibited many serious needs for treatment services. An intensive diagnostic screening tool was administered to all RYSE clients and to the control group. This assessment showed the family pattern of young women who had experienced extensive exposure to violence in the formative years. Many of the girls had family members who were in prisons and jails. A very large proportion of the young women had been victimized sexually, often by relatives and family friends. There were serious indications of mental health problems and substance abuse.

Although the girls expressed interest in school, they were not enrolled in many school-based programs such as sports, music and arts, leadership clubs, or other extracurricular activities. Despite often tragic histories of family maltreatment, these young women sought to reconnect to family members and to create positive and loving family environments. They wanted to learn and find gainful employment, but they lacked the personal resources to achieve their goals.

As with many innovative juvenile justice programs, RYSE experienced a number of fits and starts on its way to full implementation. In the main, however, the program met many of the hopes of its designers. One of the most interesting developments was the concern expressed by the RYSE probation staff that they were overwhelmed by their caseloads. This seemed a bit hard to understand given that their workloads were less than one third of those of their fellow officers. It was learned that the girls-only caseloads confronted the probation officers with a number of clients with enormous needs, who could be simply ignored within a larger caseload of boys and girls. The probation officers had multiple interactions with their troubled clients each week and could not avoid knowing about a range of very serious problems, including potential suicide attempts. Not much older than their clients, the probation officers called for more clinical resources for advice in dealing with individual cases. Once this issue was properly understood by probation leadership, the clinical resources were identified and made available.

The immediate outcome results of RYSE were encouraging. During the time that the young women were receiving intensive services and attention, the RYSE clients had a lower recidivism rate than the control group. Further, even when violations of the **conditions of probation** occurred, the court was more inclined to keep girls in RYSE, whereas it was more likely to escalate the sanctions for girls on regular probation. Thus, there was some evidence that RYSE did divert girls from detention or out-of-home placements.

Longer-term outcome results were less positive. In the 12-month period in which the RYSE clients were transitioned back to conventional probation supervision or to no supervision, the failure rates of the RYSE clients were quite similar to those of the control group. This suggests that RYSE was not able to inoculate the girls from the adverse risk factors in their young lives, nor was RYSE able to substantially change those risk factors. This finding is virtually identical to research on a number of innovative juvenile sanction programs. Benefits are clearly observable when services are being delivered in an intensive manner, but the effects decline as these services are withdrawn. One lesson may be that services to at-risk adolescents must be sustained for a much longer time than has been previously assumed. Overcoming years of maltreatment and neglect cannot be accomplished in a narrow time frame.

After the state funding for RYSE was reduced, Alameda County decided to continue a less intensive version of the program for all girls assigned to probation. The Sister Friends curriculum has been shortened and perfected. Caseloads are larger than under the original RYSE program, but county officials continue to believe that gender-specific services should be a key component of juvenile probation services. Following the launching of RYSE, several other counties have attempted to replicate the program or develop their own versions of gender-specific probation services. The results of research on some of these programs are still pending.

OTHER PROMISING GENDER-SPECIFIC APPROACHES

To date, most of the innovative gender-specific programs have focused on prevention and early intervention services. The best-known national program is operated by Girls Incorporated (formerly Girls Clubs of America), which annually serves more than 350,000 in 900-plus program locations (Weiss et al., 1996). Serving young women aged 7 to 18 years, Girls Inc., through its affiliates, offers a broad range of educational programs and peer group support. The Girls Inc. program in Alameda County has paid special attention to conflict resolution and violence prevention curricula. In Harrisburg, Pennsylvania, Girls Inc. has pioneered an innovative approach aimed at assisting at-risk girls to learn business entrepreneurship skills and to improve their job skills. The employment and business curriculum is connected to other Girls Inc. areas including reproductive health, sports, and peer education.

The Girl Scouts Behind Bars program serves adult women in custody and their daughters aged 5 to 18 years. Children whose parents are incarcerated are at very high risk of delinquent behavior, and there are very few services or programs for these children (Acoca & Austin, 1996; Mann, 1984). Volunteers from the Girl Scouts of America help the young girls to visit their incarcerated parents and maintain positive relationships. The children are also introduced to the beneficial aspects of traditional Girl Scouts programming. Incarcerated mothers and their daughters participate together in scouts activities.

Very promising community-based programs for young women involved in the justice system include the Female Intervention Team (FIT) in Baltimore, Maryland, which provided a catalyst for starting the RYSE program. The FIT program attempts to work with young women in ways that emphasize personal strength and building resiliency. There is a Rites of Passage curriculum, and girls celebrate their accomplishments at a graduation ceremony. The Kalamazoo County, Michigan, juvenile court operates a female offender program that offers wilderness and other physical challenge experiences to court-involved girls. The program uses many community-based groups and volunteers to deliver its services.

Perhaps the best known agency for gender-specific services is the PACE Center for Girls, Inc., which offers a range of educationally focused, community-based services throughout Florida. PACE operates alternative schools and supports local juvenile courts with high-quality counseling and educational services.

Increasingly, there are high-quality gender-specific residential programs such as the Harriet Tubman Residential Center in New York, the Marycrest Euphrasia Center in Ohio, the Touchstone Program in Connecticut, and First Avenue West in Iowa. Each of these programs is attempting to evolve an effective treatment response for very troubled young women that is developmentally appropriate and draws from the best professional wisdom on gender-specific services (Weiss et al., 1996).

None of the programs listed in this section have been subject to adequate research. Although promising, we simply cannot say for sure that their programs are truly effective. Further, the state-of-the-art research on girls' programming is so lacking that we cannot be certain that the intervention models that are employed by these programs are at all useful.

CONCLUSIONS

The sorry state of knowledge and innovative programming for at-risk girls cannot be allowed to continue. Although young women may continue to represent a smaller percentage than boys in the juvenile justice system, we ignore their plight at our own risk. These troubled young women are being drawn into the justice system in larger numbers. Moreover, often young parents themselves, some are locked into re-creating the cycle of violence and abuse that affects the entire society.

Simple fairness demands that young women not be subjected to excessive social control merely to protect them from themselves, even as we ignore similar behavior for young males. We cannot tolerate an inferior justice system response for girls, even as we acknowledge that the system for boys is grossly inadequate as well. There needs to be an adequate supply of gender-specific services and programs so that girls are not routed into detention and other out-of-home placements when home-based services would be more cost effective. Finally, we need to build effective community continua of responses that reduce violence and sexual abuse against young women in their homes, in their schools and neighborhoods, and in the justice system.

SUMMARY

Young women make up approximately 29% of all of the arrests of persons under age 18, but they account for a much larger share of arrests for running away and prostitution. While there has been a slight decline in arrests for girls in the past 10 years, the overall rate of young women entering the juvenile justice system has increased. Zero-tolerance policies in schools and more aggressive police responses to family disputes have adversely affected young women. Research on those young women who enter the juvenile justice system suggests that they often have histories of physical and sexual abuse. Girls in the juvenile justice system have severe problems with substance abuse and mental health issues.

There are very few juvenile justice programs that are specifically designed for young women. **Gender-responsive** programs and policies are urgently needed. Further, there is concern that young women are not getting equitable treatment by the juvenile justice system. Girls are often detained or placed in residential programs "for their own protection." The juvenile justice system is likely to incarcerate some young women for minor lawbreaking behavior, such as juvenile status offenses, that would not result in the confinement of young men. Once in the juvenile justice system, girls experience substandard health care, poor education programs, and few staff that are trained in female adolescent development. Even worse, there are well-documented instances in which young women have been sexually abused by juvenile justice personnel.

A few promising gender-responsive programs have been attempted, although not many of them have been rigorously evaluated. The RYSE program in Alameda County, California, is one example of a gender-specific program that was tested and showed some success compared to conventional methods. The OJJDP is paying increased attention to girls in the justice system and has committed to funding new research and gender-responsive program innovations.

REVIEW QUESTIONS

1. What are the crimes for which young women are arrested? What are recent trends in terms of young women and juvenile delinquency?

2. What is bootstrapping, and how has this played a role in the growing numbers of young women entering the juvenile justice system?

3. How are girls treated by juvenile corrections facilities?

4. What are the core principles that should guide the design of more gender-responsive juvenile justice programs?

INTERNET RESOURCES

The Delores Barr Weaver Policy Center is a major center for research, demonstration projects, and technical assistance to help girls in the juvenile justice system.
http://www.seethegirl.org

The Georgetown Center on Poverty and Inequality conducts research, offers consultations, provides intensive training for juvenile court personnel, and has a focus on girls in the juvenile court system.
https://www.law.georgetown.edu/academics/centers-institutes/poverty-inequality/

The Women's Foundation is a private foundation that funds research and programs designed to improve the treatment of girls and women in the justice system.
https://www.thewomensfoundation.org/

Chapter 7

Is There a Science of Prevention?

The Politics of Prevention

During the highly partisan debate in 1993 on the federal crime bill, congressional Republicans seized on the theme that delinquency prevention was little more than political pork and that this funding was little more than a payoff to selected jurisdictions in exchange for votes to support the crime bill package. An Oakland, California, program that was designed to divert high-risk youths from the streets in the evening, Midnight Basketball, was especially singled out as a worthless expenditure of taxpayer dollars. President Bill Clinton had proposed that there be a large expenditure of funding for youth development programs. Further, Vice President Al Gore championed a federal Ounce of Prevention Council that would distribute these funds and conduct additional research to discover new prevention strategies. The political battle was quite heated.

There was an important sense in which political considerations did loom very large in the debate over crime-prevention funding. Traditional Democrats, especially the Black Congressional Caucus, were opposed to the provisions of the crime bill that expanded the federal death penalty and toughened such sentencing provisions as the Three Strikes and You're Out policy that vastly lengthened sentences for repeat offenders. Liberals were told to take the funding for prevention to placate their constituents so that the president could convince more conservative members of Congress to support his bill. Ultimately, funding for the prevention programs was diminished. The ambitious Ounce of Prevention Council was given less than $1 million, while billions were allocated for more police, additional prison construction, and correctional boot camps.

This was not the first time that funding for prevention had become a political battleground. As early as the 1950s, Congress held hearings on the rising tide of juvenile crime and considered serious proposals to fund prevention programs on a national scale. President Eisenhower vetoed one proposal, arguing that crime control was a local responsibility. This changed rapidly during the John F. Kennedy administration. Attorney General Robert F. Kennedy set up the federal Office of Juvenile Delinquency that began testing community-wide prevention efforts. These early efforts were enlarged as a core component of President Lyndon B. Johnson's War on Poverty. Throughout the Richard Nixon administration, significant federal funding for positive youth development was earmarked to fight youth crime. It was in 1974, during Gerald Ford's brief presidency, that Congress created the Office of Juvenile Justice and Delinquency Prevention (OJJDP) to establish a concerted federal effort to encourage prevention of youth crime and establish community-based responses. In the wake of the Watergate scandals, President Ford, a nonelected chief executive, signed OJJDP into law despite his reservations about the value of prevention, but proposed minimal funding for the new agency. President Jimmy Carter later increased support for prevention programs through substantially expanded funding of OJJDP—the budget grew 20-fold in the first years of the Carter administration. Neither Presidents Ronald Reagan nor George Bush supported prevention funding. On the contrary, they both tried to eliminate OJJDP throughout their administrations.

Limited prevention funding was continued through strong congressional support during the Reagan and Bush years. There was a modest increase in prevention funding under the Clinton administration that favored block grants to states to determine investments in prevention programming. President George W. Bush and an emerging conservative majority in the U.S. Congress advocated that funding for prevention programming and youth development activities be reduced dramatically (Krisberg & Vuong, 2009). One exception was the new emphasis of funding faith-based organizations to lead prevention programming in areas of drug abuse, sexual abstinence, truancy, and juvenile crime. President Obama was very sympathetic to funding early prevention programs, but he inherited financial disaster that rocked the United States in 2008. Deep concerns over a rising federal deficit made claims for expanded funding for youth and family programs and scuttled efforts to rebuild the infrastructure of youth development programming.

BUILDING A FOUNDATION FOR PREVENTION PROGRAMMING

Throughout those years, many academic critics bemoaned the poor scientific foundations of most prevention efforts. Many others echoed the tough judgment of distinguished criminologist Peter Lejins (1967):

> The field of prevention is by far the least developed area in criminology. Current popular views are naive, vague, most erroneous, and mostly devoid of any research findings; there is a demand for action based on bygone days and other equally invalid reasons. In scientific and professional circles the subject of prevention has received remarkably little attention. Even the basic concepts in the field of prevention lack precision. There has been very little theory-building, and attempted research under these circumstances has failed to produce any significant results. (p. 1)

As OJJDP started national research and demonstration efforts on youth development programs, the claims that these efforts would impact serious youth crime were questioned. Through the National Assessment of Prevention, Walker, Cardarelli, and Billingsley (1976) conducted a literature search on existing programs, finding that few programs possessed a coherent theoretical foundation. Even worse, most prevention programs were so badly designed that there were significant logical inconsistencies in their program elements—client selection criteria, services, and disconnected stated outcome measures. Another national survey of delinquency prevention programs, funded by the National Science Foundation, reached a similarly dismal conclusion about the state of the art based on a review of evaluation studies (Wright & Dixon, 1977).

The first major national OJJDP initiative attempted to demonstrate that positive youth development activities offered to low-income youths by established "character-building agencies" such as the Boys and Girls Clubs of America, the YMCA, the Salvation Army, and the United Neighborhood Centers of America would divert youngsters from careers in youth crime. Many of the grantees, especially the national organizations, had been very active in lobbying Congress for the establishment of OJJDP. Some viewed this program somewhat cynically, as a payback to loyal supporters.

More than $20 million was distributed to a total of 168 agencies across the nation. The national evaluation of the initiative estimated that nearly 20,000 youngsters received some services as part of this effort. The evaluator faced innumerable obstacles. Few of these agencies could differentiate between their ordinary services and those activities that were targeted to prevent delinquency. Data collection on clients was not part of regular agency operations. There were very few chances to implement even minimally rigorous research designs. Further, many program staff resented the very concept of accountability. One often heard the opinion that "saving one youth is worth a million dollars," so more quantitative assessments of impact were viewed as inappropriate.

The first OJJDP national prevention program produced few measurable results. The vast majority of clients were no different from previous recipients of agency services. These clients typically had very little contact with the juvenile justice system; they were the "good kids." Services were not well documented. It was hard to determine how many youths received which services or the intensity and duration of these services. These agencies often met with other character-building agencies in their communities but had few ongoing relationships with schools, law enforcement, or social service agencies. There was little doubt that many of the grantees preferred to work with the most amenable youngsters, often subtly excluding those at highest risk of delinquency. In sum, the OJJDP grantees mirrored the full range of problems that had been identified by previous critics of delinquency prevention programs. Although it is hard to argue with the goal of providing healthy development experiences for all children, it is extremely questionable whether these efforts can seriously impact rates of youth crime (Krisberg, 1981).

In the aftermath of the disappointing results of the first OJJDP foray into the area of delinquency prevention, there was some effort to launch a more rigorous theory-based prevention program—the Seattle Social Development Project (Hawkins & Weis, 1985). The recommendation for the Seattle experiment emerged from a year-long literature search conducted by the OJJDP Assessment Center on Delinquent Behavior and Prevention.

The ambitious Seattle Social Development Project attempted to operationalize the precepts of delinquency theory, implement an experimental design, and follow youngsters over a 10-year period. However, the general lack of interest at the federal level in crime prevention during the 1980s doomed the Seattle project in its formative years.

Others argued for a different approach to prevention that emphasized comprehensive planning (National Advisory Committee on Criminal Justice Standards and Goals, 1976). Although this approach recognized the critical role of the systematic analysis of youth development programs and the identification of service gaps, there were few resources to help communities build or fund their plans.

Beginning in the early 1990s, some of the OJJDP leadership embraced a very different approach to delinquency prevention. A series of large-scale longitudinal studies began in Pittsburgh, Pennsylvania; Denver, Colorado; and Rochester, New York, as part of the comprehensive Program of Research on the Causes and Correlates of Delinquency (Thornberry, Huizinga, & Loeber, 1995). The goal of this research program was to discover empirical regularities in the initiation, persistence, and desistence of delinquent behavior. Further, the parallel studies would test common hypotheses, use similar measurements, and permit replication across sites. The OJJDP research staff expected that this major investment in basic research would yield substantial insight to guide effective prevention and intervention programs.

The Program of Research on the Causes and Correlates of Delinquency assumed that sound data could generate useful ideas for policy makers and practitioners. The program established the concept that there are "developmental pathways to delinquency" that could be discerned among preadolescents and that these might change as the youth matured (Kelley, Thornberry, & Smith, 1997). The researchers argued for a comprehensive community-wide approach that might involve multiple aspects of a child's life; they said different services would be more or less important at different ages. This was a significant departure from the single factor (and single agency) approach that had held sway in the past among youth development practitioners. OJJDP sponsored a number of subsequent study groups to produce ongoing assessments of research findings that could be used to guide national programming (Loeber & Farrington, 1998, 2001).

IMPLEMENTING EFFECTIVE COMMUNITY PREVENTION MODELS

In 1992, Congress enacted Title V as an amendment to the federal Juvenile Justice Act. Title V, also known as the Community Prevention Grants Program, was championed by local elected officials and youth advocates as a mechanism to direct limited federal funds to effective prevention programs. In fiscal year 1994, the Title V program allocated $13 million to local communities that participated in successful comprehensive planning efforts. The Community Prevention Grants Program was then funded at $20 million annually in fiscal year 1995 through 1998, $45 million in 1999, and $42.5 million in both 2000 and 2001. To assist localities, OJJDP launched a training and technical assistance program to help communities meet the mandates of Title V. This provided both a challenge and a strategic opportunity for federal officials to translate the findings of the Causes and Correlates program, as well as other federally sponsored research projects, into practical guidance for youth workers and local elected officials.

The OJJDP utilized the academic and community implementation experience that was building at the University of Washington. The work of David Hawkins and Richard Catalano (1992) was well suited to bridge the worlds of research and prevention practice. Drawing heavily on concepts that were gaining currency in the health field, Hawkins and Catalano incorporated the foundations of health promotion into their model. They pointed out that most adults know about the primary risk factors for cardiac disease such as genetic predisposition, obesity, smoking, and lack of exercise. They also know about protective factors—steps to take to lower the probability of heart trouble even given several risk factors. Many Americans began taking greater note of their own risk profiles for heart disease and took action to increase their protective factors. This greater awareness of risk and protective factors, according to Hawkins and Catalano, resulted in a major decline in deaths from cardiac disease.

By analogy, delinquency prevention could be conceptualized as a process of ameliorating risk factors and increasing protective factors that are directly related to adolescent misconduct. Hawkins and Catalano developed a very sophisticated community planning model to accomplish these objectives. The model, based on the social development strategy and titled Communities That Care, was utilized by OJJDP in its Title V training efforts.

Communities That Care derived its insights on risk factors from a range of longitudinal studies of adolescent development. It offered trainees an easy-to-comprehend framework through which a large body of research could be summarized and communities could organize their prevention efforts. Table 7.1 shows the risk factors incorporated into the Communities That Care model. They are organized along the four domains of community, family, school, and individual and peers.

Table 7.1 covers five troublesome adolescent behaviors: substance abuse, delinquency, teenage pregnancy, dropping out of school, and violence. For a risk factor to get a check mark or be connected to a particular problem behavior, there had to be at least two rigorous studies linking that risk factor with the problem behavior. The absence of a check does not mean that there is no relationship but simply that the research is not conclusive.

Hawkins and Catalano chose not to assign statistical weights to the various risk factors to make it easier for nonresearchers to utilize this information. They did show that as more risks were present in a youth's life, the odds of engaging in the problem behavior increased geometrically (Hawkins & Catalano, 1992). The fact that some risk factors appear across multiple types of problem behavior illustrates how these behaviors are interrelated. For example, extreme economic deprivation, family management problems, family conflict, early and persistent antisocial behavior, and early academic failure are powerful risk factors for virtually all types of adolescent problem behavior. Moreover, reductions in some risk factors can have a multiplier effect across many areas of undesirable youth behavior.

It is worth noting that these risk factors operate at both the individual and community levels. In particular, the community risk factors are primarily attributes of neighborhoods in which at-risk families and youths live. The linkage between these risk factors and individuals is often mediated by protective factors, which is discussed shortly. By contrast, the school risk factors are descriptive of negative individual youth experiences, although different school environments may produce more of these risk factors.

Protective factors are derived from the social development strategy and a theory of delinquent behavior that enjoys substantial empirical support (Howell, 1995). The social development strategy suggests that young people can be insulated from risk factors by holding healthy beliefs and engaging in **prosocial** behaviors. Youngsters acquire these positive beliefs and clear standards by becoming bonded with individuals, community institutions, and peer groups that hold these views. The more that young people interact with and are surrounded by prosocial people, the less chance that they will be attracted by negative and antisocial groups such as youth gangs.

The bonding process occurs when young people can actively participate in prosocial groups. It is here that young people acquire knowledge and skills, are given opportunities to demonstrate these new skills, and receive strong positive recognition, a sense of belonging, and public rewards for their behavior. Hawkins often illustrates the power of the bonding process by making an analogy to delinquent gangs. These negative peer groups permit young people to acquire a wide range of criminal knowledge and skills—how to buy and sell drugs, how to use guns, or how to avoid apprehension by law enforcement. Gangs offer emotionally powerful recognition, which is especially meaningful for vulnerable and isolated young people. New gang members can demonstrate their new skills by acquiring drugs for the gang, attacking gang rivals, or being a lookout during robberies and burglaries. Gangs offer a sense of belonging that is vital to alienated young people who often refer to their gang as a family. Delinquent peer groups offer praise and tangible rewards (money and illegal goods and services). Seen from this vantage, a gang can make very effective

Table 7.1   Adolescent Problem Behaviors

Risk Factor	Substance Abuse	Delinquency	Teenage Pregnancy	School Dropout	Violence
Community					
Availability of drugs	✓				
Availability of firearms		✓			✓
Community laws and norms favorable toward drug use, firearms, and crime	✓	✓			✓
Media portrayals of violence					✓
Transitions and mobility	✓	✓		✓	
Low neighborhood attachment and community organization	✓	✓			✓
Extreme economic deprivation	✓	✓	✓	✓	✓
Family					
Family history of the problem behavior	✓	✓	✓	✓	
Family management problems	✓	✓	✓	✓	✓
Family conflict	✓	✓	✓	✓	✓
Favorable parental attitudes and involvement in the problem behavior	✓	✓			✓
School					
Early and persistent antisocial behavior	✓	✓	✓	✓	✓
Academic failure beginning in elementary school	✓	✓	✓	✓	✓
Lack of commitment to school	✓	✓	✓	✓	
Individual/Peer					
Rebelliousness	✓	✓		✓	
Friends who engage in the problem behavior	✓	✓	✓	✓	✓
Favorable attitudes toward the problem behavior	✓	✓	✓	✓	
Early initiation of the problem	✓	✓	✓	✓	✓
Constitutional factors	✓	✓			✓

SOURCE: Wikipedia/Preventionbetterthancure

CASE STUDY: IS THERE A SCIENCE OF PREVENTION?

Daniel was raised in a very violent neighborhood. His family experienced severe economic hardships and struggled to get the basic necessities of life. His mom become quite ill and could not work. Daniel had great difficulties in high school and had failing grades. He had no hopes for completing his education, was extremely depressed, and considered dropping out of school.

A school counselor suggested that he apply for a paid summer internship in which students were tutored and learned the basics of law. He participated in moot court competitions and worked part time in a district attorney's office. Daniel explained that the program built up his self-confidence and proved to him that he was smart.

He completed the summer program and later returned as a mentor for others. He has graduated from college and completed a law degree from Vanderbilt University and clerked for a U.S. Attorney. He now works for the state bar association to expand the number of underserved minorities in the legal profession.

use of the tenets of the social development strategy. The problem, however, is that young people are being bonded to individuals who hold unhealthy and procriminal beliefs and attitudes.

The social development strategy is suggestive of a broad range of positive learning experiences including music, the visual and theater arts, sports, and other areas that are very attractive to young people. These opportunities for acquiring knowledge and skills, participating, and being rewarded can be provided by families, faith-based groups, schools, and any number of community-based youth groups. Connecting young people to these positive development experiences is the primary means of protecting against risk factors. Hawkins and his colleagues have reported on research demonstrating that increasing the number and improving the quality of protective factors can substantially reduce the adverse effects of multiple risk factors. However, this research shows that sustained exposure to a large number of risk factors can overwhelm the long-term value of increasing protective factors. Thus, Hawkins and Catalano urge communities to work simultaneously on reducing the range of risk factors as they increase the number and extent of protective factors. A one-sided approach to prevention will not succeed (Hawkins & Catalano, 1992).

Communities That Care helps communities conduct self-inventories and assessments of the broad range of risk and protective factors present in their environments. Local key leaders are coached to measure their communities against others using archival data or youth surveys. Based on these data, the leaders are asked to prioritize a limited set of prevention priorities that will receive primary attention. Once these areas are identified, training is given on research-based programs that can reduce the selected risk factors and increase protective factors. Communities are encouraged to establish multiyear plans to implement their localized prevention strategies. The Communities That Care model is a major departure from the traditional approach to prevention that emphasized initiating more programs without regard to whether these programs were effective or genuinely responded to local needs. The traditional failed method of prevention programming is to (a) latch onto media-created panaceas (the Scared Straight or Tough Love programs are two such examples), (b) dispense funding to programs that have no proven impact but have powerful backers (Drug Abuse Resistance Education is an example of a politically favorable program with no positive impact on participants), or (c) distribute public dollars to community groups that are adept at garnering political support through their well-connected board members or alumni (for example, Boys and Girls Clubs of America).

Research on the usefulness of the Communities That Care model through the Title V program was very encouraging in that it allowed both researchers and the communities themselves to understand which methods worked and why. This also benefited OJJDP by identifying which strategies were most efficient in preventing juvenile delinquency (Chibnall, 2001; Hsia, 2001). A subsequent OJJDP program that melded the risk-focused prevention approach of Communities That Care with research-based reforms of the juvenile sanctioning system produced similar and very positive results (Krisberg, Barry, & Sharrock, 2004).

A different perspective on prevention has been advanced by the advocates of an *assets* or *strength-based* approach to youth development (Benson, 1997). Proponents of this view assert that too much attention has been paid to weaknesses and problems, some of which cannot be easily changed. The goal of this approach is to recognize strengths and assets that exist in otherwise impoverished families and communities. One goal is to build on existing community resources in lieu of saturating communities with outside treatment professionals. Another key objective is to develop the resiliency of young people facing difficult life situations (Leffert et al., 1998). Whereas the research foundation for Communities That Care is largely dependent on large-scale empirical studies that incorporate longitudinal designs, the research support for the assets or strengths model is derived from the clinical insights of psychologists or social workers. These latter studies are usually based on a small number of case studies and qualitative research. The Communities That Care approach draws more heavily on sociological theory and statistical analyses.

In several communities, these two approaches to community-wide prevention efforts competed for public attention. There are for-profit organizations that are marketing training packages in these different approaches, as well as a range of products for parents, schools, and communities. Not infrequently, different governmental agencies will select one or the other approach, leading to heated battles over which perspective will predominate. The diffuse nature of government responsibility for funding and providing youth development services across multiple county and municipal departments makes turf competition endemic to the prevention field. In fact, the two broad approaches to delinquency prevention can be used to complement each other.

Further complicating the local discussion on juvenile crime prevention is the seemingly endless array of individual programs or curricula that claim to prevent troublesome adolescent behavior. Few of these prevention solutions have any empirical evidence to support their claims. A literal cottage industry exists, producing booklets, video and audio tapes, board games, and computer exercises that purport to prevent school violence, teenage pregnancy, drug abuse, tobacco use, and bullying. Often, the same materials or programs are modestly changed to focus on the adolescent social problem that gains current media attention and attracts public funding. For example, after the tragic shootings at a number of suburban and rural high schools, there was a sudden upsurge in the availability of allegedly effective school programs and curricula to prevent school killings. As the media began painting a picture of the offenders as victims of bullies, the prevention field sprang into action to market antibullying programs. Another media-driven area of prevention programs is the growth of private boot camps that promise worried parents that the in-your-face approach can head off a complex of risky adolescent behaviors including substance abuse, promiscuity, rebelliousness, and poor school grades. There is not a shred of rigorous evidence supporting their claims. In fact, the research on the presumed effectiveness of government-financed juvenile boot camps has been largely negative (MacKenzie, Grover, Armstrong, & Mitchell, 2001).

A particularly noteworthy prevention program was operated by a small police department in northern California. Called Elvis and the Lawmen, the program consists of a small number of law enforcement officers who visit local elementary and middle schools dressed up like Elvis Presley and perform some of the King's greatest hits. After the performance, Elvis and his friends speak about the consequences of drugs, violence, and smoking or try to promote traffic safety—depending on the latest funding stream. The group has its own newsletter that is distributed statewide and attracts significant funding from public sources and private contributions. The effectiveness of the Elvis and the Lawmen program has never been evaluated. Running a close second to the Elvis program was a well-funded program jointly sponsored by Hand Guns Incorporated and the National Basketball Association. In this program, the staff, sometimes accompanied by professional basketball players, work with children in kindergarten and preschool. The program culminated when the 5- and 6-year-old children signed a pledge that they would not engage in violence or play with guns. This is hardly an effective long-term method for deterring delinquency. Another school antiviolence program funded by the U.S. Department of Education employed jugglers dressed as clowns to assist with school conflicts. The list of superfluous prevention programs is quite long, lending credibility to the claims of critics that prevention is pork.

An encouraging development in recent years has been the dissemination of a series of publications that report on prevention programs that have been vetted by solid research. For example, Lawrence Sherman and his colleagues at the University of Maryland (Sherman et al., 1997) produced a comprehensive evaluation for the National Institute of Justice of the research support for particular prevention programs that were sponsored by the federal government. A series of more targeted research reviews was funded by OJJDP and formed the basis of a number of community training efforts across the nation (Howell, Krisberg, Hawkins, & Wilson, 1995; Lipsey & Wilson, 1998). The work of Elliott (Alexander, Pugh, & Parsons, 1998) through the Blueprints for Violence Prevention project is especially noteworthy. A research team at the University of Colorado identified a limited number of prevention programs that had been carefully and rigorously evaluated. But beyond the usual brief description of successful programs, Elliott and his associates provided in-depth information on how they were structured, initiated, and maintained. A series of publications from the Blueprints project is an important resource for those communities genuinely interested in implementing proven approaches.

Farrington and Welsh (2007) have recently assembled a comprehensive listing of individual-focused delinquency prevention programs. They identified several studies that concluded that enriched preschool intellectual environments, home visiting programs with troubled families, and training parents in appropriate discipline practices have been shown to be effective in reducing future offending behavior. School-based programs that help teachers better manage disruptive behavior in classrooms as well as improved teaching and learning strategies have been shown to reduce delinquency. Moreover, expanded after-school programs and community-based options to engage youth in positive activities with prosocial adults appear to be effective in reducing troubling youth behaviors. There is support for the value of curriculum that uses cognitive behavioral methods to strengthen the ability of young people to increase their self-control, especially to avoid violent situations, and to increase the capacity of at-risk youth to upgrade their abilities to communicate and to interact with adults. There are several studies supporting the value of well-designed mentoring programs such as those run by Big Brothers and Big Sisters. There is also tentative evidence that arts, music, and cultural programs can be helpful in insulating at-risk youth from negative peer and neighborhood influences.

INTERRUPTING THE CYCLE OF VIOLENCE

Rigorous research on the lives of chronic and serious juvenile offenders and a number of highly encouraging program evaluations direct our attention to prevention efforts that focus on children caught up in the tragic world of family violence. Several cross-sectional studies have examined the prevalence of abuse and neglect among offender populations. For example, Pawasarat (1991) found that 66% of male offenders and 39% of female offenders in Wisconsin were victims of substantiated reports of abuse and neglect. Another study in three states of high-risk juveniles on parole who had allegedly been victims of abuse and neglect ranged from 29% in Virginia to 45% in Colorado and to 53% in Nevada (Osofsky, 2001). Moreover, several longitudinal studies have demonstrated the powerful developmental effects of child abuse and neglect (Kelley et al., 1997; Widom, 1992, 1995; Wordes & Nuñez, 2002).

Figure 7.1 from the work of Kelley and her associates (1997) illustrates that maltreatment in childhood is associated with much higher rates of violence, pregnancy, drug abuse, poor school grades, and mental health problems. For instance, 70% of child victims of abuse reported engaging in violent behavior as teenagers compared to 56% of nonabused children. Childhood abuse victims were more likely to have an official record of delinquency (45% versus 32%) and were twice as likely to have multiple problems during adolescence (Kelley et al., 1997).

Widom (1992) followed a cohort of 1,575 children through their adolescent years and into adulthood. Maltreated children were more likely to be arrested as juveniles (27% compared

Figure 7.1   Relationship Between Prevalence of Child Maltreatment and Various Outcomes During Adolescence , Kelley et al. (1997, p. 8).

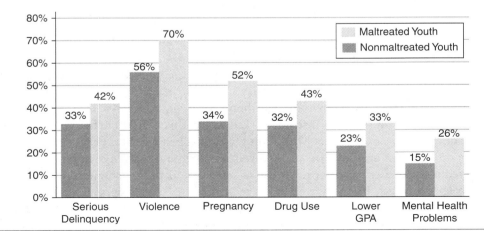

SOURCE: Kelley et al. (1997, p. 8). U.S. Department of Justice, Office of Juvenile Justice and Delinquency Prevention.

to 17% for nonmaltreated children). The two groups also differed in terms of probability of adult arrests (42% versus 33%). Moreover, the maltreated children had a higher average number of adult arrests (5.7 compared to 4.2). Child victims of abuse and neglect also were likely to commit violent crimes as teenagers and adults.

America suffers from a serious epidemic of abuse and neglect of its children. While no one knows for sure how many children are maltreated, the nation's child protective service (CPS) agencies received official reports on approximately 3 million allegedly maltreated children in 1999. Of these reported cases, 28% were substantiated after investigations that led to some formal state intervention with these families (U.S. Department of Health and Human Services, 1999). Analyzing only the substantiated cases of maltreatment, children age 3 or younger had the highest rate of neglect, and boys ages 4–11 had the highest rates of physical abuse. Girls were most likely to be sexually abused, with 12- to 15-year-olds at the highest risk for sexual abuse. Children younger than 6 accounted for 86% of all fatalities due to child abuse and neglect, with half of this group being infants less than 1 year (Osofsky, 2001).

In addition to being the direct victims of violence, children also witness violence in their homes. Compared to other children, youth who witness domestic violence (whether or not they are victims themselves) are more likely to exhibit a wide variety of behavioral problems, more symptoms of anxiety and depression, and more aggressive conduct (Kolbo, Blakely, & Engleman, 1996). Further, several studies suggest that there is a significant overlap between cases involving violence against women and violence against children in these same families. For example, English (1997) found that 55% of referrals to child protective services for physical and emotional abuse involved children who were present during episodes of domestic violence. A study that examined hospital medical records of 116 children who were suspected of being abused or neglected found that 45% of the mothers had histories of partner abuse. When a similar analysis was conducted one year later, the researchers found that 59% of the mothers of abused children had medical records suggesting that they had been battered by their partners (U.S. Department of Health and Human Services, 1999). It is very clear from these studies that America's hidden victims of domestic terrorism, those who live in daily fear of domestic violence, comprise a very substantial segment of serious and chronic juvenile offenders. Further, there are several studies suggesting that poor parenting techniques involving inconsistent or overly harsh discipline or excessively frequent family conflicts (those that might stop short of battering) are causal factors for delinquency (Thornberry, 1994).

These research studies of child abuse and domestic violence suggest that targeted prevention programs aimed at these vulnerable children should produce substantial youth crime control impacts. There are several tested programmatic approaches that appear to be helpful in reducing the risks of family violence (Developmental Research and Programs, 2000). These prevention efforts can range from marital therapy to improved methods of responding to reports of child maltreatment.

Studies by Hahlweg and colleagues (1998) report that communication difficulties among couples before marriage and before conflicts arise are strong predictors of escalation in distress among couples. Marital counseling programs that teach effective communication skills and problem-solving and conflict resolution techniques can be meaningful prevention programs. The best programs allow couples to learn and practice these skills with the assistance of a therapist or counselor. One such tested program is the Prevention and Relationship Enrichment Program (PREP), which involves approximately 15 hours of lectures, videotapes, and role-playing exercises. PREP couples report a range of positive outcomes, including fewer instances of physical violence with their spouses, compared to couples in control groups (Markman, Renick, Floyd, Stanley, & Clements, 1993).

Prenatal programs and programs for infants have also proven their capacity to recognize the early predictors of maltreatment and offer meaningful prevention services (Barnard, Booth, Mitchell, & Telzrow, 1988; Horachek, Ramey, Campbell, Hoffman, & Fletcher, 1987; Olds & Kitzman, 1993). Often, these programs are delivered by medical professionals as part of regular prenatal medical care that covers education for the mother on physiological and emotional stresses of pregnancy, preparing for childbirth, and guidance on promoting the mental and physical well-being of newborns. The conversations and coaching in the doctor's office are supplemented by periodic home visits by nurses or other health professionals. For example, the Prenatal/Early Pregnancy Program developed by David Olds and his colleagues (Olds, Henderson, Chamberlin, & Tatelbaum, 1986; Olds & Kitzman, 1993) has proven that it can reduce tobacco, alcohol, and narcotics use among expectant mothers, thus significantly reducing perinatal childhood health problems. Participants in this program evidenced a 4% abuse and neglect rate compared with 19% for a control group in the 2 years after program involvement. A subsequent follow-up study that tracked program participants for the next 15 years showed fewer signs of adolescent misconduct such as running away, substance abuse, and delinquency (Olds, Hill, & Rumsey, 1998).

There are several other tested programs that can avert or lessen the impact of family violence, for example, functional family therapy (FFT; Alexander et al., 1998). In this model, a well-trained therapist conducts home visits with a family for 12 to 36 hours over a 90-day period. During these visits, generally 1 hour in duration, the therapist takes the family through a series of key developmental steps. First, the therapist engages the family and builds motivation for change. This process often involves breaking down ingrained negativity, increasing respect among family members for differing values and beliefs, and rebuilding trusting relationships. The FFT worker conducts a careful family assessment in which characteristics of family members are identified. This assessment should result in referrals to intensive services to respond to issues such as addiction, anger management, health, and mental health concerns that are underlying family problems. The therapist works to improve parental skills through teaching, specific family assignments, and other training techniques. The FFT worker deals with a broad range of internal and external family needs including interactions with schools, welfare agencies, the extended family, and employers. Through the Blueprints project, Elliott and his colleagues (Alexander et al., 1998) located 13 articles in high-quality scientific journals that document the effectiveness of functional family therapy. Some of these studies followed families for up to 5 years and have reported reductions of offending by the children of at least 25% or as much as 60% compared with control group children.

Other early interventions with maltreated children, or those who have witnessed domestic violence, include the Child Development-Community Policing (CD-CP) Program that was jointly conducted by the New Haven Police Department and the Yale University Child Study Center (Marans et al., 1995). This program forms teams of police and child development specialists who work together on behalf of physically or emotionally traumatized children. Besides joint training that equips the police with a greater ability to recognize the impacts of trauma on children, the CD-CP program provides on-call clinicians who immediately minister to children that are part of crime scenes involving violence. To date, there are several clinical observations that support the value of this effort (Marans et al., 1995). (There is ongoing research and evaluation on the project through the Child Study Center, the Northeast Program Evaluation Center at the Yale School of Medicine Department of Psychiatry, and the Department of Psychiatry Center for

Health Care Services Research at the University of Connecticut Health Center, but no published results.)

Besides implementing a comprehensive continuum of *programs* responding to child treatment, effective prevention requires a *research-based decision-making system* that ensures that high-risk children and families get appropriate attention. This latter goal poses a major problem for many communities. CPS agencies are often inundated with reports of child abuse that are well beyond their ability to respond to due to limited staff resources. It is not uncommon for CPS agencies to investigate only 60% of the reports that they receive. Some urbanized jurisdictions do not even meet this frequency of investigations. In Michigan, for example, approximately 11% of reports of child abuse and neglect were fully investigated (Baird, Wagner, Healy, & Johnson, 1999). Further, the rate of worker turnover is quite high at these agencies, leaving the burden of often life or death decisions to inexperienced and untrained staff (Freitag & Wordes, 2001). Moreover, many CPS agencies depend on fairly unstructured methods requiring extensive individual worker judgments that may be affected by racial, ethnic, or social class biases. These *clinical decision-making systems* have been successfully challenged in federal courts as arbitrary and capricious (Baird et al., 1999).

Several locales are implementing structured decision-making (SDM) systems that are more effective in identifying high-risk cases and directing priority legal and social service attention to these cases. SDM systems, now in place in cities in Michigan, California, Wisconsin, Minnesota, and New Hampshire, among other states, utilize research-based risk and needs assessment tools to determine how cases should be handled. SDM agencies initiate staff training that outlines the presumptive actions that the worker should take based on the results of the research-based assessment tools. Workers need to consult with their immediate supervisors if they want to override these presumptive actions. Research completed in Michigan, and now underway in California, confirms the assumptions of the developers of SDM that rates of re-abuse would be lower, that fewer serious and fatal injuries would occur in SDM locales, and that treatment resources would be targeted more effectively (Baird, Wagner, Caskey, & Neuenfeldt, 1995). As noted earlier, communities must have a full array of services to prevent chronic child maltreatment and effective methods for identifying families that would benefit from these early interventions.

THE COST-EFFECTIVENESS OF PREVENTION

Several researchers have tried to quantify the relative costs and benefits associated with prevention programs compared to other crime-control policies. For instance, Yoshikawa (1994, 1995) examines the costs and benefits associated with nearly 40 parent-training programs. He identified four programs that appeared to reduce long-term effects of serious and chronic delinquency. A highly influential study by the RAND Corporation identified another set of prevention efforts that appeared to produce substantial crime reduction impacts for relatively modest amounts of program investments (Greenwood, Model, Rydell, & Chiesa, 1996). RAND researchers focused on home visits to at-risk families by child care professionals, training for parents with very young children who were acting out in school, cash and other incentives to encourage low-income high school students to graduate, and close monitoring of high school–age youths who were showing early signs of problem behavior. The RAND report cautions the reader that the cost and benefit estimates are based on the results of pilot studies and that results seen after taking these programs to scale may be somewhat different. Greenwood and his associates found that the graduation incentives save enough in averted crime costs to pay for the entire program.

Parent training and intensive supervision of minor delinquents would save between 20% and 40% of program costs. The study concluded that home visits would be an effective but costly program. As noted earlier in this chapter, the use of research-based assessment tools to target the highest risk families could increase the cost-effectiveness of family visitation programs.

The RAND analysis also compared the effectiveness of prevention programs to a California law that significantly increased penalties for offenders with three felony convictions (the Three Strikes and You're Out provision). Greenwood and his colleagues (1994) had previously estimated that full implementation of Three Strikes could reduce crime in California by 21%. However, the same Peter Greenwood later concludes that prevention programs could reduce state crime rates by up to 15% at radically lower costs to the taxpayer. Figure 7.2 suggests that Three

Figure 7.2   Cost-Effectiveness of Early Interventions, Compared With That of California's Three-Strikes Law

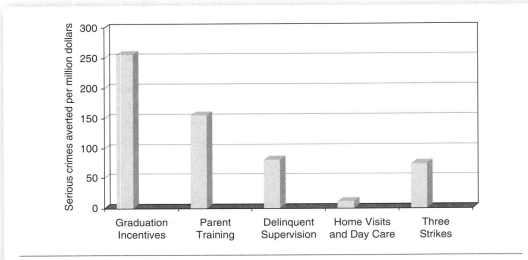

SOURCE: From Greenwood et al., *Three Strikes and You're Out: Estimated Benefits and Costs of California's New Mandatory Sentencing Law.* Reprinted with permission of RAND.

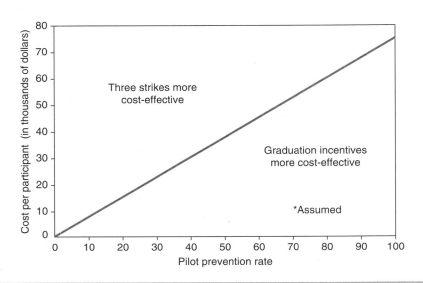

Strikes would cost California $5.5 billion to implement but that a statewide investment of about $1 billion in early prevention programs would equal the crime reduction benefits of the Three Strikes policy (Greenwood et al., 1996). Greenwood and his associates called for more demonstration projects involving prevention programs to develop better estimates of the true potential of early intervention and prevention efforts.

Researchers from the state of Washington have also tried to identify cost-effective prevention programs through a sophisticated economic model (Aos, Phipps, & Korinek, 1999). These researchers caution us that not all prevention programs are effective in averting future offenses. Although a range of prevention programs were found to be effective, the Washington group emphasized that success rates were modest—even the most successful program lowered rates of adult offending by only 10% to 15% (Aos et al., 1999). The research review suggested that some prevention programs do work for certain youths in certain contextual settings.

The Washington study found that programs for juvenile offenders could be quite cost effective in terms of reducing future rates of adult criminality. Aos and his colleagues (1999) suggest that the research base of effective prevention strategies is still nascent and requires further development, including well-designed replications. Aos and his associates summarize their findings across a wide variety of interventions. Cost savings were computed for both criminal justice system costs and costs to victims. Intensive home-based programs for juvenile offenders produce the greatest cost savings per dollar expended. One reason for this finding is that the juvenile justice programs are targeted at youths who are already showing extremely high probability of continued and serious offending. The early childhood education and family support programs appear to be less cost effective. This result may be a direct by-product of policies that offer child development programs to all low-income families without identifying the small proportion of these families that will contain future serious and chronic offenders. More focused delivery of services to the highest risk families would likely increase the bottom line, or monetary benefit, for these programs. Of course, early-childhood programs are meant to accomplish multiple social objectives beyond crime control, and this must be factored into these economic calculations.

There is some support for the positive effects of community-wide prevention interventions. Some cities have tried to utilize reformed adult offenders to help divert troubled youth from criminal behavior. These programs also aim at reaching families through neighborhood activities, feeding very poor youth, and expanding community policing models. An excellent example of this approach that tries to counteract a community subculture of violence is the Ceasefire program designed by public health officials at the University of Illinois, Chicago (Skogan, Hartnett, Bump, & Dubois, 2009).

IS THERE A SCIENCE OF PREVENTION?

There are still too many community prevention efforts that do appear to be pork—political payoffs by legislators for local constituency groups. The state of the art in prevention still needs further refinement. Sadly, like Gresham's Law of Currency, cheap and worthless prevention programs tend to drive out the "hard" or effective programs. However, there is little doubt that a beginning science of prevention has emerged. Modern prevention programming is more likely to be guided by careful research on the etiology of youth crime and by rigorous evaluations of prevention programs. To the extent that political and public policy support grows for implementing knowledge-based prevention programming, society may see smarter crime control investment strategies. Interestingly, some law enforcement groups—the National Crime Prevention Council and Fight Crime, Invest in Children—have joined in the conversation about the best prevention approaches to increase the safety of our communities.

SUMMARY

Efforts to prevent juvenile delinquency have often been hotly debated in political venues. Liberals tend to like prevention programs, and conservatives often view them as useless social programs. The federal government's funding of delinquency prevention began during the Kennedy administration and continued to the passage of the Juvenile Justice and Delinquency Prevention Act (JJDPA) of 1974. The initial prevention programs supported by OJJDP were not well designed: Selecting appropriate clients was one issue, and documenting the services that were provided was another major concern. Traditionally, prevention programs have taken the "cream," the most amenable youth, to work with, leaving out those needy youngsters from more difficult family backgrounds.

At the request of youth advocates and other groups, Congress amended the JJDPA in 1992 to include Title V, which directed some funding to localities to support prevention programs. OJJDP also provided high-quality training on the best practices in the prevention arena and funded a number of significant research studies and special study groups that expanded our understanding about the causes and correlates of serious delinquent behavior. The work of David Hawkins and Richard Catalano and their model approach to prevention, Communities That Care, greatly advanced the state of the art in research-tested prevention programming.

Increasingly, studies are showing the critical relationships between young people who are victims of or witnesses to violence in the home and involvement in serious and violent crime. A focus on the most vulnerable and maltreated young people could make better use of available prevention funding. There are many proven programs that can be helpful for these families, including premarital counseling, visiting nurses for high-risk families, comprehensive early childhood education, and intensive family counseling programs.

REVIEW QUESTIONS

1. What are risk and protective factors, and how can they be utilized to plan and organize prevention efforts?

2. What are the factors that have slowed the development of effective prevention programs?

3. What are the differences between the prevention models advocated by the administrations of presidents Kennedy, Johnson, Reagan, Clinton, and Obama?

4. What are the characteristics of prevention programs that have been found to be the most effective?

INTERNET RESOURCES

Youth.gov is a comprehensive resource for information about prevention programming for youth that is compiled across many federal agencies including the DOE, NIMH, CDC, DHS, and the DOJ.
http://www.youth.gov/

Crime Solutions is a federally promulgated website that offers listings of research-vetted prevention programs.
http://www.crimesolutions.gov/

The Prevention Institute is a leading nonprofit that offers training and technical assistance to improve community-based prevention programs.
https://www.preventioninstitute.org

The California Endowment Foundation offers funding, training, and seminars for organizations wishing to expand prevention programming in a variety of health fields, including delinquency and drug abuse.
http://www.calendow.org

Chapter 8

WHAT WORKS IN JUVENILE JUSTICE

In the 1970s and 1980s, it was common for academic observers and policy makers to conclude that "nothing works" in juvenile justice (Martinson, 1974). Conservative critics of juvenile justice argued that rehabilitation programs were not effective and that greater emphasis must be placed on deterrence and incapacitation (Murray & Cox, 1979; Wilson, 1983). Critics of juvenile justice from the left claimed that juvenile justice programs actually stigmatized youths, making them worse than if they had not been "helped" (Lemert, 1951; Schur, 1973). Conservatives claimed that the rehabilitative ideal of the juvenile court meant that its sanctions were too lenient, teaching youngsters that they could violate the law with impunity. Liberals felt that juvenile sanctions were often too harsh in relationship to the generally less serious crimes committed by young people. They pointed out that young women, the poor, children, and young people of color disproportionately felt the brunt of these overly harsh punishments.

Lack of confidence in the efficacy of juvenile court sanctions helped fuel a national movement to toughen penalties for juveniles, including new laws making it far easier to try minors in criminal courts (Torbet et al., 1996). The *get tough* lobby felt that the new laws would produce more punishment. At least some civil libertarians (Feld, 1984) argued that the criminal courts offered greater safeguards in terms of due process and equal protection of law.

Neither ideological side of this public policy debate paid much attention to a growing body of research suggesting that many juvenile justice sanctions were, in fact, effective. For example, careful studies of individual programs beginning in the 1960s showed many examples of very positive results. Research by Palmer (1971), employing a classic experimental design, demonstrated that intensive community treatment programs (in which parole officers had no more than 12 clients) produced a 1-year completion of parole success rate of 82%, compared to a 65% success rate for youths sent to the California Youth Authority. After 2 years, the intensive community treatment group had a 61% success rate, compared to 40% for the controls. A later reanalysis of Palmer's data by Lerman (1975) suggests that these results were a bit exaggerated but that the far less expensive community treatment produced no worse results than traditional youth incarceration policies. A second randomized experiment, conducted in Los Angeles, also found that community-based treatment was as effective as incarceration but at less than half the cost (Empey & Lubeck, 1971). Other studies in Utah by Empey and Erickson (1972) and in Illinois by Murray and Cox (1979) demonstrated that very intensive probation or parole supervision could significantly reduce the incidence of reoffending. However, not all studies of intensive parole (or aftercare services) produced such positive results. A study in Pittsburgh, Pennsylvania, and Detroit, Michigan, found little difference in the results of youths receiving intensive aftercare services compared to a control group receiving traditional parole supervision (Greenwood & Turner, 1993). The authors of this study suggest that the lack of statistically reliable differences might be attributable to the low sample sizes in the experimental and control groups.

Besides a number of studies of discrete programs, there have been careful examinations of entire state systems that emphasize alternatives to traditional youth confinement policies. In the early 1970s, the state of Massachusetts closed all of its traditional juvenile correctional facilities (these are often referred to as "state training schools"), moving nearly 1,000 youths to a varied network of small secure placements, community-based group homes, and intensive home-based services. This development was especially dramatic because it involved closing the first state training school in America, the Thomas Lyman School. For some, the Massachusetts juvenile correctional reforms constituted one of the most dramatic changes ever in juvenile justice policy (Miller, 1998).

Remarkably, the Massachusetts reform movement has been extremely durable. Since the early 1970s, the state has never opened a large juvenile correctional facility, despite a two-decade national policy debate that emphasized increasing use of incarceration for juveniles. Secure facilities are reserved for the most serious offenders, which comprise only about 15% of all commitments to the Massachusetts Department of Youth Services (DYS). The largest secure facility in Massachusetts holds less than 50 youths, as compared to juvenile correctional facilities in other states that house hundreds of juvenile offenders. The basic philosophy and organizational structures that helped close the Massachusetts state facilities remain largely in place today. Interestingly, despite some periodic debates about juvenile justice policy in Massachusetts, most Bay State residents are not even aware of how radically different their system of youth corrections appears when compared to other jurisdictions.

The Massachusetts reforms were first studied by a team of Harvard University researchers headed up by Professor Lloyd Ohlin (Coates, Miller, & Ohlin, 1978). The Harvard researchers compared the outcomes of youngsters released from the new network of community-based programs in 1974 to the results observed with those who were released from the state facilities in 1968, before the reforms were implemented. These initial research results were not encouraging. The recidivism rate after 12 months for the postreform cohort was actually higher than for the prereform group (74% versus 66%). One possible explanation for these negative results was that less serious offenders were being diverted from the state system during the postreform period, thus creating a cohort of youths who were likely to be more chronic and serious offenders compared to earlier cohorts. Ohlin and colleagues also noted that the recidivism data differed in various regions of Massachusetts. They noted that in areas of the state that had done a more thorough job of implementing the new community-based system of corrections, the recidivism rates were equal to or better than those of the prereform cohort (Coates et al., 1978).

With the support of the Edna McConnell Clark Foundation, the National Council on Crime and Delinquency (NCCD) completed a second research report on the Massachusetts community-based corrections system in 1989. One goal of this study was to see if a more mature and settled set of reforms was producing better outcomes. In fact, this *is* what NCCD found. The recidivism rate for those released in 1989 was 51%—much better than the results from either the prereform days or the early 1970s (Krisberg, Austin, & Steele, 1991). Further, multiple measures of postrelease failure showed that Massachusetts youths performed as well as, and often better than, youths from juvenile correctional systems in other states. The NCCD study of Massachusetts found a significant decline in the incidence and severity of DYS clients committing offenses when the 12 months prior to their admission to DYS was compared to the 12-month period after they returned to the community. These reductions in law-violating behavior were sustained over the next 2 years. NCCD also reported that the Massachusetts reforms were cost-effective, saving the state more than $11 million per year compared to more conventional juvenile correctional policies.

A related study by NCCD researchers of the juvenile corrections system in Utah produced equally encouraging results (Krisberg, Austin, Joe, & Steele, 1988). In the wake of serious scandals at the state training school, Utah had adopted many of the same reforms as Massachusetts in the late 1970s. Although the overall failure rates for Utah youths were higher than those for youngsters in Massachusetts, these youths showed a major decline in the severity and rate of offending after participation in community-based programs and detention in small secure facilities. Other states such as Missouri, Maryland, New Jersey, Pennsylvania, and Vermont attempted to implement some or all of the elements of the Massachusetts approach to juvenile corrections. Only Pennsylvania and Maryland conducted data-based evaluations of these efforts. The research by Goodstein and Sontheimer (1987), which covered 10 community-based residential programs in Pennsylvania, had reported outcomes that were equivalent to the NCCD study of Massachusetts. However, Gottfredson and Barton (1992) found less promising outcomes after the closing of a major state training school in Maryland. In their study, youths who were committed to the Maryland Division of Youth Services and placed in community-based programs were compared with youths who had been released from the Montrose State Training School before the reforms. The Montrose clients actually did better in terms of recidivism than those in the community programs. These Maryland results are reminiscent of the Harvard study in Massachusetts. It may well be the case that it takes time to implement effective systems of community-based corrections and that results measured in the early months of a reform will be disappointing.

The state of Ohio is another important locale for research on effective juvenile sanctioning programs. Under the title of Reclaim Ohio, the state youth corrections agency developed a strategy in which the most violent youths were housed in state-run correctional facilities, but Ohio counties were given funding to strengthen local programs for less serious offenders. A similar program had been attempted in California in the 1970s, producing mixed results (Lemert & Dill, 1978; Lerman, 1975; Palmer, 1992). Before the implementation of Reclaim Ohio, the state had one of the highest juvenile incarceration rates in the nation. Proponents of the new program observed that the counties were able to send unlimited numbers of juvenile offenders to state correctional facilities at no budgetary cost to local units of government.

Under the Reclaim program, counties were given funding that was proportionate to the number of juvenile felony adjudications, but they were now responsible for paying the incarceration costs associated with youths from their jurisdiction. Youthful offenders who were convicted of the most serious crimes would still be taken by the Ohio Department of Youth Services (ODYS) without costs to the localities. Counties could use the money to develop local programming options. Nine Ohio counties were pilot sites in 1994, with the program going statewide in 1996. Researchers at the University of Cincinnati evaluated the results of the original nine counties to determine if there had been a genuine reduction of commitments and if local juvenile justice officials were able to launch effective juvenile justice sanctions. The pilot jurisdictions were viewed next to a number of comparison counties. The initial evaluation results were very supportive of Reclaim Ohio. While the pilot and control counties began with very similar rates of commitment to state facilities, the pilot counties experienced a large decline in the number of youths sent to the ODYS (a 42% decline between 1993 and 1994). By contrast, the comparison counties witnessed a 42% increase in youths sent to ODYS during this same period. Further, the researchers noted that there was not a drop in felony adjudications, suggesting that the observed results were due to a declining commitment rate rather than there being fewer offenders in the system (Latessa, Turner, & Moon, 1998). The University of Cincinnati research team reported that the reduction in commitments to ODYS comprised less serious offenders—the group that was targeted for diversion by the Reclaim program. They also reported that the pilot counties were able to retain roughly 42% of their fiscal allocations under the Reclaim programs. Each pilot county began at least one new program, and there appeared to be stronger networks of juvenile sanctioning programs in most of the pilot sites contrasted with the comparison counties (Latessa et al., 1998).

As noted earlier, over 40 states downsized their incarcerated youth populations until about 1990, but the effort to close juvenile facilities ran up against a renewed hostility toward young offenders who were defined as "superpredators." The number of incarcerated young people rose in the late 1990s and generated a new wave of litigation against juvenile corrections (Krisberg, Marchionna, & Hartney, 2015). Most states responded to the public awareness about abusive practices and conditions in youth prisons with new efforts to divert young people from these places, and produced another dramatic drop in the number of youngsters growing up behind walls. This second wave of decarceration was strongly influenced by developments in the state of Missouri.

Today, the best-known policies to replicate the closure of secure training schools were implemented by the state of Missouri. The Show Me State closed its only large training school and moved youth to small regional facilities and several innovative home-based programs. Following the leadership of the famed researcher Jerome Miller, Missouri created small facilities that looked more like college dorms than youth prisons, and they allowed youth to wear normal clothing, not prison garb. One of the remarkable aspects of the Missouri reforms was that they survived the "get tough" blowback of the 1990s, and they were supported by conservative and liberal governors alike. While there has yet to be rigorous outcome research on the Missouri approach, this approach to juvenile corrections was actively disseminated by the Annie E. Casey Foundation, and the Missouri model was partially adopted by many states (Mendel, 2011, 2014). The Missouri model helped lead to a resurgence in closing youth training schools in the first decade of the 21st century.

SUMMARIES OF PROGRAM EVALUATIONS AND META-ANALYSES

Some researchers of juvenile justice sanctioning programs have gone beyond studies of individual programs, summarizing the results of dozens of studies through sophisticated research techniques known as meta-analyses (Lipsey & Wilson, 1998). The most prominent researcher in this group is Mark Lipsey. In his work, Lipsey (1992; Lipsey & Wilson, 1998, 2001) combines multiple program evaluations and has identified intervention effects that have not been readily apparent in previous studies due to low numbers of subjects in the experimental and control groups. By statistically combining multiple studies, Lipsey has been able to document significant program results that did not emerge from single research efforts. For example, in one of his earliest meta-analyses, Lipsey examined more than 400 studies of delinquency treatment programs that incorporated experimental designs since the year 1950. In contrast to claims that most delinquency treatment programs were not effective, Lipsey found that youth in treatment groups had recidivism rates that were, on average, 10% lower than those in control groups. According to Lipsey, the best juvenile justice treatment programs showed results that were 20% to 37% better than for youths in control groups (Lipsey, 1992). The most effective programs typically involved structured training or behavioral modification programs to help youths change their ability to achieve in school, increase their job skills, and improve capacities for self-control. The most successful programs incorporated multiple therapeutic approaches and had higher levels of intensity in terms of the number of contact hours and the length of the treatment program. Lipsey found that close monitoring by researchers of the quality and integrity of the treatment programs was strongly related to positive outcomes. In general, programs that were located in the community were much more likely to show positive outcomes than programs that were operated in correctional facilities. For Lipsey, it is not a matter of asking "Does treatment of delinquency work?" but of defining the necessary conditions that maximize the effectiveness of various proven interventions.

Other researchers have reinforced and extended Lipsey's conclusions. Palmer (1996) looked at 23 literature reviews of juvenile justice and adult correctional programs, and nine meta-analyses. He, as well, concluded that behavioral approaches that emphasized skill building seem to consistently produce positive results. Similarly, cognitive treatment programs that helped offenders sort out thinking patterns that led to criminal misconduct produced generally positive results. As with Lipsey's research, Palmer found multimodal approaches that combined behavior change, cognitive development, and skill acquisition to be consistently the most promising treatment models. Palmer also reported that programs that attempted to confront or threaten offenders (e.g., programs such as Scared Straight that used prisoners to frighten delinquents) were not at all effective. Traditional casework techniques were not found effective in multiple studies. Moreover, a variety of family interventions, vocational training, employment programs, or wilderness challenge programs met with, at best, mixed results (Palmer, 1996).

Several summative examinations of the most effective juvenile justice programs have highlighted the critical ingredients of these programs. For instance, Altschuler and Armstrong (1984) point to six critical components of successful programs for juvenile offenders. They argue that the best programs use "continuous case management" in which one staff person monitors a youth's progress toward certain well-defined treatment goals from the earliest stages of program entry into the period after program completion. Continuous case management also rests on the ability of staff to offer a broad range of highly individualized services that are closely tied to the young person's needs. Altschuler and Armstrong believe that good programs must pay special attention to community reentry and reintegration for youths who have been placed out of their homes. Also important are programmatic components that permit youths to participate in making program decisions, and programs that offer clear opportunities for youths to demonstrate achievement of knowledge and skills. The best interventions offer enriched educational and vocational content. The reader may note that these latter program components are quite similar to the findings reported in Chapter 7 about the attributes of the most successful prevention programs. Altschuler and Armstrong (1984) make a strong case for the importance of juvenile justice sanctioning programs having clear and consistently applied consequences for youths who violate program rules.

Strikingly similar to the conclusions of Altschuler and Armstrong, Peter Greenwood and Franklin Zimring (1985) have offered their own list of core program ingredients based on an independent review of the delinquency treatment literature. They pointed to the value of providing young people with opportunities to develop positive self-images, reducing the influence of negative role models, and creating ways to bond youths to prosocial adults and social institutions. Greenwood and Zimring also emphasized the importance of allowing youths to discuss their childhood problems, especially their early exposure to violence and abuse. Similar to other program summaries and meta-analyses, Greenwood and Zimring echoed the value of cognitive programs that assisted youngsters in comprehending harmful thought patterns, particularly the ways in which criminal behavior was rationalized. For Greenwood and Zimring, effective programs must offer a range of services that are closely matched to particular youth needs. A broad review of delinquency programs completed for the National Institute of Justice by Lawrence Sherman and his colleagues (1997) at the University of Maryland and a comprehensive review by researchers Gendreau, Little, and Goggin (1996) further confirm the key program ingredients cited by other researchers.

A GRADUATED SYSTEM OF SANCTIONS AND INTERVENTIONS

The studies discussed above have formed the foundation of modern thinking on how juvenile justice systems ought to be structured. The Office of Juvenile Justice and Delinquency Prevention (OJJDP) supported several research efforts that examined the elements of effective juvenile justice programs. The basic concepts supporting the new idea of a graduated system of sanctions and interventions were developed in connection with OJJDP's Comprehensive Strategy for Serious, Violent, and Chronic Offenders (Howell, 1995). In 2002, OJJDP commenced a multiyear training and technical assistance effort with NCCD and the National Council of Juvenile and Family Court Judges to expand the implementation of model systems of **graduated sanctions**. Although there does not exist a perfect real-world model of the most effective system of graduated sanctions, many jurisdictions have made significant progress in this direction. Further, policy makers are increasingly demanding that practitioners incorporate this best thinking in new budget and program proposals.

Too often, juvenile justice officials have put forward a single program as the panacea for curing juvenile offenders. Piled on this trash heap of failed juvenile justice wonder programs have been Scared Straight, reduced caseloads for probation and parole officers, wilderness challenge programs, and boot camps—all discredited based on careful and sound evaluations. Large-scale juvenile training schools have long been shown to be ineffective in reducing recidivism, yet advocates for institutions have constantly tried to repackage their programs using the latest psychological fads such as Tough Love, Reality Therapy, Victim Awareness, Transcendental Meditation, Positive Peer Culture, and so on. While there are clearly some program interventions with proven and research-based outcomes, it is not really plausible that one single therapeutic or programmatic approach will be effective for the wide range of needs of the young people who come into the juvenile justice system.

The concept of a graduated system of sanctions and interventions directs our attention toward an array of services and programs that should be available to troubled youths. Key to this notion is the idea of a continuum of responses that range from relatively unobtrusive and low-cost interventions for first-time or very minor offenders to more structured and comprehensive interventions for more serious and chronic juvenile offenders. The underlying theory of graduated sanctions is that the juvenile justice system needs to provide a flexible range of sanctions and interventions. As a youth's behavior gets worse, the system should be capable of increasing the level and intensity of supervision and services. If the youngster demonstrates appropriate and desired behavioral change, the juvenile justice system needs to be able to ratchet down the level of control so that the youth can move toward an independent and law-abiding life. Think of the analogy of a deep-sea diver. If the diver is at a very low depth under water and attempts to rise to the surface too quickly, he or she will suffer a severe and painful condition known as the *bends*. If, however, the diver comes to the surface in a more gradual manner (often remaining at certain depths for short periods of time), the effects of the changed levels of pressure are mitigated. Unfortunately, the current juvenile justice system ignores or fails to respond to many instances of youth misconduct, often because of perceived limits on existing correctional resources. Then, the juvenile court will impose very restrictive sanctions such as long-term placements in secure training schools. Youths may spend many months in a facility in which virtually every aspect of daily life (i.e., wake-up times; meals; recreational hours; personal hygiene; drug, alcohol, and tobacco consumption; and bed times) is rigidly controlled by institutional rules. Upon release, these same youths are now required to reenter the real world in which they must confront many choices, some of which propel them back to criminal behavior. One way of explaining the very high rates of failure for young people who are released from secure facilities is that they suffer from a social and psychological equivalent of the bends. Legal scholar Franklin Zimring (in press) suggests that the juvenile justice system may be thought of as a "muscle-bound giant." The giant is very slow to move, and when it moves, it does too much. Systems of graduated sanctions and interventions are designed to fix these major flaws in most juvenile justice systems.

The continuum of graduated sanctions should include immediate, intermediate, and secure sanctions (Wilson & Howell, 1993). Immediate sanctions are intended for youths who have committed relatively minor offenses or have come to the attention of the juvenile court for the first time. These interventions communicate to young people that there are clear consequences for misconduct. Immediate sanction programs are often targeted at juvenile status offenses such as running away, truancy, curfew violations, or chronic conflict with parents. These programs have been shown to be effective in preventing the escalation of youthful problem behavior to more serious law violations. Although immediate sanctions need not be expensive or of extended duration, these early juvenile justice interventions provide an opportunity to recognize risk factors that might lead to further problem behavior. One well-documented immediate sanction program is the Michigan State Diversion (MSD) Project. The Michigan effort has been subjected to rigorous experimental testing and has shown consistently lower recidivism rates for its clients compared to youths who were released with no services or supervision (Davidson et al., 1977; Davidson, Redner, Blakely, Mitchell, & Emshoff, 1987). The MSD project assumes that minor offenders are best handled outside the formal justice system by working with family and community resources. The program uses college students to work with the MSD clients. The college students receive eight weeks of training in behavioral change and client advocacy. The college students seek to direct youths to needed services as well as to provide one-on-one contacts in the community. Caseloads are very small, rarely exceeding 10 cases. Other immediate sanction programs that have demonstrated the value of intensive, nonjustice system interventions include the Choice Program in Maryland (Maton, Seifert, & Zapert, 1991) and the North Carolina Intensive Protection Supervision Project (Land, Land, & Williams, 1990).

Intermediate sanctions are designed for youths whose behavior has escalated to serious law violations. These youths engage in chronic misconduct, often involving drug use and property crimes. In some cases, intermediate sanctions can be used for youngsters who have committed very serious offenses but have never before come to the attention of legal authorities. These

---

CASE STUDY: WHAT WORKS IN JUVENILE JUSTICE

A seventh grader in an Albuquerque middle school was clowning in class by making loud burps. His teacher reported this 13-year-old to a school-based police officer who searched the boy for drug possession. A school vice principal accused this youth of selling drugs to other students. The youth was asked to remove his shoes and pants and the officer searched his underwear. No drugs were found in this search, but the matter did not end there. The boy was moved to a juvenile detention center and later suspended for the rest of the school year. The boy was charged with interfering with the education process by threatening or inciting others to obstruct the valid mission or functions of the school. The public officials saw this relatively common benign school misbehavior as indicative of a society in moral decline. A U.S. Court of Appeals upheld the actions of the school and police officials since they "believed" that they had no other options.

---

youths are not appropriate for placements in secure correctional facilities, although they are often sent there due to the lack of meaningful intermediate sanction programs. These programs provide a full range of very close supervision of the youth, intensive work with families, and a rich mixture of treatment services. These are costly programs, but they often cost less than one fifth of out-of-home placement or secure confinement (Henggeler, Melton, & Smith, 1992). One of the most well-researched intermediate sanction programs is multisystemic therapy (MST), which has demonstrated its effectiveness in treating serious and chronic offenders in a number of different settings (Borduin et al., 1995; Family Services Research Center, 1995). MST views troubled youths as "nested within a complex of interconnected systems that encompass individual, family, and extra familial (peer, school) factors; and interventions may be necessary in any one or a combination of these systems" (Family Services Research Center, 1995, p. 5).

MST works with youths in the context of their home and community environments. Based in part on the successes that family preservation services in the child welfare arena have had in avoiding foster-care placements, MST offers very intensive home-based services. MST workers may spend as much as 60 contact hours over a 4-month period. The frequency of therapist contact usually tapers off during the last few weeks. The MST therapist is very well trained and is assigned no more than four to six families at any given time. One should contrast this level of services with traditional juvenile probation programs in which the youth might see a probation officer once per month, and this meeting would typically occur in the probation office, not in the home. A typical probation officer might have a minimum of 50 clients, or as many as 100 youngsters, to monitor on a regular basis. Probation officers receive minimal or no training in working with families; they are primarily focused on ascertaining if the youth has obeyed the court's orders.

Other intervention programs that have proven their effectiveness include functional family therapy (Alexander, Pugh, & Parsons, 1998) and the Family and Neighborhood Services program based on the principles of MST (Henggeler et al., 1992). A number of programs have successfully intensified probation services by increasing the frequency of client–probation officer contacts and enriching treatment services available to youths on probation. These include the Lucas County Intensive Supervision Unit (Wiebush, 1993) and the Wayne County Intensive Probation Program

(Barton & Butts, 1988). Intermediate sanctions also include day treatment programs in which the youth spends the day at an educationally based treatment program and returns home at night. These latter programs are especially helpful for youngsters who are experiencing great difficulties in reentering traditional schools. At the upper end of intermediate sanctions are residential facilities that are nonsecure. These facilities may be located in both urban and rural areas, and youths may leave the facilities during the day to attend school or to perform community service. A small, nonsecure residential program, the Thomas O'Farrell Youth Center (TOYC), is located in a rural/suburban area outside of Baltimore, Maryland, and uses the *normative model* to reform its youthful wards. The normative model assumes that learning and helping others adhere to positive group values and expectations is crucial for teaching troubled adolescents how to make appropriate life choices. In addition to high-quality clinical and educational services, TOYC utilizes an intensive peer group process to achieve its goals. The average stay at TOYC is 9 months, with as many as 6 months of aftercare. Whereas previous research on positive peer culture has been mixed at best, studies show that TOYC has very impressive results in reducing recidivism (Krisberg, 1992). It is possible that other attempts at mobilizing peer influence as a rehabilitative tool have failed to achieve the appropriate level of intensity (at TOYC there are small groups and community meetings that occur throughout each day). Staff at TOYC are well trained in the normative model, operate under its rules, and actively participate in the group sessions.

Secure confinement programs are designed for the very small number of adolescents who have engaged in violent behavior or repetitive incidences of very serious criminal behavior. Yet even for this very difficult group of youths, there are proven programs that reduce their criminal behavior. Research has shown that locked programs must be very small in capacity; otherwise, a dangerous gang culture tends to dominate these secure facilities (Bartollas, Miller, & Dinitz, 1976; Feld, 1977). Further, secure correctional programs should be located where high-quality educational and treatment programs are available to all inmates.

In the mid-1980s, OJJDP sought to develop a treatment model for violent juvenile offenders. The Violent Juvenile Offender (VJO) program was tested in Boston, Massachusetts; Detroit, Michigan; Memphis, Tennessee; and Newark, New Jersey (Mathias, Demuro, & Allinson, 1984). Program clients were initially placed in small locked facilities but were gradually phased into community-based residential facilities and then supervised very closely when they returned home. In VJO, youths had been adjudicated for a violent offense and had at least one prior adjudication for a very serious crime. The VJO clients received intensive treatment services and were subject to continuous case management from their initial entry to secure confinement until their release into the community. Youths were given tangible opportunities for prosocial achievements and were quickly sanctioned for violating program rules. Implementation of the VJO model was reasonably successful in Boston and Detroit but was never fully operational in Memphis or Newark. The VJO program evaluation employed a classic experimental design comparing the program to more traditional incarceration practices at the four test sites. The research found that there were no significant results in the sites that poorly implemented the VJO approach, but VJO youths were arrested less frequently and for less serious offenses than control groups in Boston and Detroit (Fagan, 1990).

The Florida Environmental Institute (FEI) was another example of an innovative secure sanction program. Operated by Associated Marine Institutes, the FEI was reserved for youngsters who would otherwise be sent to Florida's adult prisons. The average stay in FEI was 18 months, and almost all residents return to their home communities upon release. While FEI is not a locked facility, the remote location in the Florida Everglades, which is surrounded by swamps and dense vegetation, makes escape unlikely. Besides the desired attention to education and treatment needs, FEI engages its troubled clients in environmental reclamation work. There is a close staff-to-client ratio of no more than one to four. Despite small sample sizes and the need for more rigorous research designs, research to date has shown FEI to yield very promising results (Krisberg, Currie, Onek, & Wiebush, 1995).

Perhaps the most impressive secure confinement program is the Capital Offender Program (COP) of the Texas Youth Commission (TYC), which is reserved for young people committed for homicide. Youths are sent to a state training school for an average incarceration period of up to 3 years. During this stay, the juvenile murderers are involved in a 16-week small-group program with eight youths and two or three staff members who live together in a separate living unit for the duration of the intensive program. Youths in the COP have already been at the training school for 1 year and have at least 6 months left to serve. The COP is designed neither for youths with severe mental health problems nor for those with significant developmental disabilities.

Group psychotherapy and intensive role playing form the core of the COP intervention. Youths confront their own often abuse-filled growing-up experiences. Role playing covers dramatic reenactment of the homicide from the perspective both of the offender and of the victim and asks the offender to imagine the experience of the crime from the point of view of the victim. The program is run by highly trained clinical staff with extensive experience in psychotherapy and small-group work.

Research sponsored by the Texas Youth Commission has documented measurable personality changes in the COP youths. Program participants showed decreased levels of aggression and anger and increased ability to exert self-control and to accept personal responsibility for past actions. The COP clients showed improved ability to express empathy for the suffering of their victims. TYC researchers collected data on arrests and convictions of COP participants for up to 3 years after their release from the training school. These data were compared with a control group of capital offenders who did not experience the COP program. The treatment group showed a lower arrest rate than that for the controls (22% versus 40%) after a 1-year follow-up period. Longer-term follow-up differences between COP and control youths were less impressive, suggesting that to improve overall program outcomes, a longer period of more intensive and phased aftercare services should be considered.

The VJO, FEI, and COP outcome results are impressive in that they were accomplished with those youngsters for whom policy makers and the public-at-large are most likely to demand long-term confinement in adult prisons. Combined with the very positive results emerging from well-designed immediate and intermediate sanction programs, the foregoing research suggests that meaningful continua of graduated sanctions can be implemented at more reasonable costs and with no compromise of public safety. Moving juvenile justice systems in this direction can free up public funds to expand support for prevention programming.

Figure 8.1 summarizes what has been learned about the core components of highly effective graduated sanction programs. The lower right section of Figure 8.1 restates the basic notion that there must exist a full continuum of interventions with a range of immediate, intermediate, and secure sanctioning options, with aftercare services needed to support each discrete sanctioning program.

The lower left side of Figure 8.1 lists treatment and services that have proven to be effective with seriously delinquent youths. These programs can exist as part of the various components of the continuum of sanctions. The top of Figure 8.1 refers to knowledge, skills, tools, and practices that form the infrastructure of a fully effective graduated system.

Figure 8.1   Components of a Model System of Graduated Sanctions

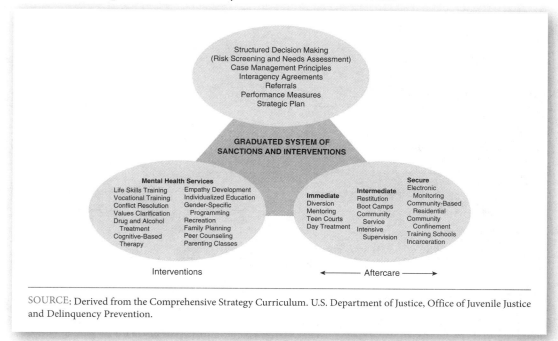

SOURCE: Derived from the Comprehensive Strategy Curriculum. U.S. Department of Justice, Office of Juvenile Justice and Delinquency Prevention.

There is no more important part of this continuum of sanctions than the use of structured decision-making tools. These consist of research-based assessment or screening instruments to permit juvenile justice decisions to be made on objective criteria that are directly related to protecting public safety. Typically, objective risk screening tools utilize a small number of factors that are given numerical weights. See Form 8.1 for an example of such an instrument.

The higher the score, the more likely that the individual will reoffend. These instruments do not claim to provide perfect predictions of future behavior, which is nearly impossible. The instruments place youths into distinct risk pools that possess very different probabilities of further criminal conduct. Once the juvenile justice worker scores the individual case, there are presumptive expectations about how the youth will be assigned to a range of graduated sanctions. The worker may override the presumed action, but this generally requires that an immediate supervisor also reviews the case.

By utilizing objective decision-making tools, the juvenile justice system can better ensure that the most dangerous cases, and those with the greatest treatment needs, are getting priority attention and supervision. Structured decision making facilitates a better allocation of scarce justice resources. On average, objective screening and assessment lead to lower levels of incarceration and appears to be fairer to minority youths (Wiebush Baird, Krisberg, & Onek, 1995). Most current juvenile justice systems rely almost exclusively on the very subjective judgments of police, detention intake workers, and probation officers. These unfettered subjective judgments mean that personal biases color how the juvenile justice system decides who will be confined and under what levels of control (Bridges & Steen, 1998). In this book, we also examine in more detail the toll of injustice that subjective decision-making systems exact on minority children and young women.

Case management refers to a process of identifying clear expectations for behavioral change that should occur within a specific sanctioning option. The case manager is required to

Form 8.1   Sample Risk Assessment Scale

	Score

1. Age at first adjudication

11 or under	3	
12–14	1	
16 or over	0	

2. Number of prior arrests

None	0	
One or two	1	
Three or more	2	

3. Current offense

Nonassaultive offense (e.g., property, drug, etc.)	2	
All others	0	

4. Number of prior out-of-home placements

One or fewer	0	
Two or more	1	

5. History of drug usage

No known use or experimentation only	0	
Regular use, serious disruption of functioning	1	

6. Current school status

Attending regularly, occasional truancy only, or graduated/GED	0	
Dropped out of school	1	
Expelled/suspended or habitually truant	2	

7. Youth was on probation at time of commitment to DSS

No	0	
Yes	1	

8. Number of runaways from prior placements

None	0	
One or more	1	

9. Number of grades behind in school

One or fewer	0	
Two or three	1	
Four or more	2	

10. Level of parental/caretaker control

Generally effective	0	
Inconsistent and/or ineffective	1	

	Score	
Little or no supervision provided	2	

11. Peer relationships

Good support and influence; associates with nondelinquent friends	0	
Not peer-oriented or some companions with delinquent orientations	2	
Most companions involved in delinquent behavior or gang involvement/membership	3	
Total Score		

Risk Assessment	0–8	Low Risk
	9–13	Moderate Risk
	14–18	High Risk

SOURCE: Howell, J. C. (1998). Youth gangs: An overview. Washington, DC: U.S. Department of Justice, Office of Justice Programs Office of Juvenile Justice and Delinquency Prevention

frequently monitor the youth's progress and to make programmatic adjustments as needed. If at all possible, the youth should have the same case manager from entry into the juvenile justice system through release from the program. Case plans need to consider the whole child, addressing health, educational, vocational, and family issues, as well as the youth's criminal conduct. The best case management plans are well documented and updated regularly. The youth and his or her family should participate in the formulation of the case plan.

Effective sanctioning systems make extensive use of referrals to other agencies to provide the full range of services that youngsters may require. These referral systems need to be formalized via interagency memoranda of understanding (MOU). Too often, juvenile justice systems depend on very informal mechanisms through which other agencies accept and provide help to delinquent youths. Unfortunately, when funds are tight, these informal agreements, which rest on worker relationships, may fall apart. Finally, sanctioning programs need to have clear rules and defined responses when youths break these rules. Young people do best when adults set well-defined and consistent boundaries, and the adults are prepared to respond with proportionate and fair restrictions to adolescents who test those limits.

Research on effective graduated sanctions programs has taken us far from superficial polarities, which conclude that every program works or that nothing works with serious and chronic juvenile offenders. We have learned that "small is beautiful"—that the best juvenile programs are of a human dimension in which staff and youths can interact in a constructive manner. The best juvenile justice programs treat the whole child and customize services and supervision to the needs of the individual youth. Moreover, effective responses to youth crime must be capable of de-escalating or escalating formal controls as a youngster's behavior improves or deteriorates. The best juvenile justice programs take advantage of the same social development process that has been proven valuable in prevention programs. Put simply, good sanctioning programs attempt to bond youths to prosocial adults by giving the youth new skills, opportunities to demonstrate those skills, and positive recognition for successful achievement. At the same time, sanctioning programs must attempt to reduce the risk factors that propel youths further into criminal conduct.

SUMMARY

The successful experiences of states such as Massachusetts, Missouri, and others have shown the elements of effective juvenile justice interventions. One core concept is the idea of a system of graduated sanctions that allows more flexibility in response to problems or behavior of young people under court supervision. What is needed is a continuum of care that offers a full range of immediate, intermediate, and secure program options. Further, the success of any placements can be enhanced with intensive and high-quality reentry services.

Placement in graduated sanctions should be guided by objective tools that measure risks of future recidivism as well as treatment needs. Juvenile justice sanctions work best if they are closely matched to youths' needs. Several studies have demonstrated the cost-effectiveness of home-based care models such as multisystemic therapy, functional family therapy, and wrap-around services. For more chronic and serious offenders, cognitive and behavioral therapies have shown very promising results.

On a systemic basis, states such as Ohio have reduced their dependence on training schools by providing more resources to communities to treat youngsters locally. Good juvenile justice programs require strong linkages between the juvenile justice system and community agencies, especially the schools. Job training and other skill-building programs have shown great promise in reducing recidivism.

REVIEW QUESTIONS

1. What were the initial research results (by the Harvard Law School) of the Massachusetts juvenile justice reforms of the 1970s? How did these results change over time, and why?

2. What are the key elements of structured decision making, and why is SDM so important in building and maintaining a model system of graduated sanctions?

3. How do the ideas of risk and protective factors and of social development theory help define the core principles of the most effective juvenile justice interventions?

4. What is the Missouri model?

5. What have various meta-analyses found in terms of the elements of effective interventions?

INTERNET RESOURCES

The Annie E. Casey Foundation is a leading children's philanthropy. The foundation sponsors research and demonstrations of innovative juvenile justice programs, including the replication of the closure of the training schools in Massachusetts and the reduction of the Juvenile Detention Alternatives Initiative.
http://www.aecf.org

Performance-based Standards is a detailed compendium of evidence-based standards to be used to improve juvenile corrections. Jurisdictions can hire the PbS group to help in the quality control of their programs.
http://www.pbstandards.org

The Council of Juvenile Correctional Administrators is a national professional association dedicated to improving juvenile correctional programs through implementation of evidence-based programs.
http://www.cjca.net

The Pew Charitable Trusts is one of the nation's largest charitable groups. This group is a major resource for government officials and community groups who want to improve the effectiveness and efficiency of justice programs. The Pew Charitable Trusts is an excellent source for trends on juvenile justice and evaluations of recommended programs.
http://www.pewtrusts.org

Chapter 9

THE GANG BUSTERS

Does Getting Tough Reduce Youth Crime?[1]

Most reviews of juvenile justice interventions have focused on the effectiveness of prevention and alternatives to incarceration. Examinations of correctional programs have primarily covered treatment-oriented approaches. Yet the bulk of funding that is aimed at reducing juvenile crime has been devoted to programs designed to get tough with juvenile offenders. These programs are premised on several interrelated ideas, including, first, the belief that taking serious offenders off the streets will lead to crime reductions. The second underlying concept is that harsh penalties will deter other potential offenders from serious law-breaking behaviors. Third, there is the assumption that the juvenile justice system is too lenient and that minors who commit adult-type crimes should receive adult penalties. Given that get-tough strategies are expensive compared to community-based alternatives, it is reasonable to question whether research has demonstrated the efficacy of these approaches.

The move to institute more get-tough programs in the juvenile justice system rests on some research indicating that there is a small core of serious and violent offenders who account for the vast majority of serious crime committed by youth (Elliott, 1994; Wolfgang, Figlio, & Sellin, 1972). By contrast, the available data show that the caseloads of the juvenile justice system are dominated by troubled teenagers who are generally not very violent offenders. For example, about 5% of juvenile arrests are for violent crimes. There are nearly twice as many youths arrested for liquor law violations as for robbery. In 1995, 104,000 young people were arrested for vandalism compared to 2,560 who were arrested for homicide (Federal Bureau of Investigation, 1996). Approximately 8% of juvenile court referrals are for murder, robbery, assault, or forcible rape. The vast majority of the court's caseload consists of youths charged with violations of probation, minor property crimes,

---

[1]An earlier version of this chapter appeared as Barry Krisberg (1997). *The Impact of the Justice System on Serious, Violent, and Chronic Juvenile Offenders.* Oakland, CA: National Council on Crime and Delinquency.

and status offenses such as truancy, curfew violations, running away, and incorrigibility. The juvenile justice system is awash with troubled but minor offenders. However, the public policy debate on youth crime has been preoccupied with the "dangerous few" (Hamparian, 1982).

The Gang Busters

The principal target of tough law enforcement and prosecution of juvenile offenders has been members of juvenile gangs. The high-quality research on these gang suppression programs has been limited, and the findings have been inconclusive at best. A review of some of these antigang programs is instructive.

During the Reagan and Bush administrations, there was ample federal funding for gang control programs operated by police departments or prosecutors. A virtual cottage industry of gang experts emerged to offer training and assistance to law enforcement agencies. Longtime gang researcher Malcolm Klein (1995) notes that, somewhat ironically, Los Angeles was set forth as a model of gang suppression programs. However, few cities have suffered more from the violence connected with gangs than the City of Angels, despite substantial public investments in antigang programs.

Los Angeles gang control efforts have involved all aspects of the criminal and juvenile justice systems. City and county officials have tried saturation patrols, intensified surveillance, special police tactical units, and even a "ninja-style unit, complete with black clothing" (Klein, 1995, p. 1). The most widely known of these programs, called Operation Hammer, involved military-type police sweeps in the South Central area of Los Angeles. During one weekend in 1988, the Los Angeles Police Department (LAPD) sent more than 1,000 additional officers into South Central. The police assault teams made numerous arrests for traffic violations, outstanding court warrants, curfew violations, and other gang-related behavior. In all, there were more than 1,450 arrests made in one weekend. The volume of young suspects taken into custody was so great that the LAPD set up a temporary **booking** operation in the University of Southern California (USC) football stadium.

Of those arrested, all but 103 defendants were released by the police with no charges being filed. Only 32 persons were charged with felonies. The vast majority of those arrested were not gang members. The director of the California Youth Authority (CYA) noted that not one youth was sent to the state's youth corrections system as a result of the enforcement activities during Operation Hammer.

Over the next several weekends, Operation Hammer continued, although the number of police officers assigned to the antigang crusade decreased dramatically. The large volume of arrests continued, and most of these defendants were charged with outstanding warrants. There was little or no evidence that gang violence decreased in South Central or other sections of the city. The large commitment of police resources netted little in terms of the capture of illegal drugs or weapons. Operation Hammer was subjected to a careful research analysis. Few similar programs have been rigorously evaluated. An empirical review by Lawrence Sherman (1990), conducted for the National Institute of Justice, suggested that some enforcement sweeps did show short-term crime suppression results followed by a rapid decline in effectiveness. Sherman found that targeted enforcement crusades were somewhat useful in reducing parking violations, prostitution, street drug sales, and drunk driving. There is no evidence that targeted crackdowns on gangs have any effect in reducing the more serious crimes associated with these criminal groups.

Proponents of intensive police efforts such as Operation Hammer argue that these campaigns exert a deterrent effect for would-be offenders. However, research on deterrence indicates that

the swiftness and certainty of punishment is at least as important as the severity of those penalties (Zimring & Hawkins, 1973). When one evaluates Operation Hammer in light of the theory
and views the research on deterrence, it seems apparent that the program was bound to fail. The
sanctions were not swift because they were tied to outstanding warrants for offenses that had
occurred well in the past. Certainty of punishment was almost nonexistent. Few, if any, defendants that were swept up in the police raids were punished at all. Most defendants had their
charges dropped immediately. Very few alleged gang members suffered anything worse than a
brief involuntary visit to the USC football stadium. Penalties were not a core component of Operation Hammer. Indeed, many gang members and residents of South Central Los Angeles viewed
the weekend police invasions as little more than a cynical "made for television" public relations
campaign by the LAPD.

Other gang suppression programs have dispatched roaming patrol cars to "saturate" alleged
gang turfs with the goal of increasing the surveillance of gang activities and of harassing gang
members. Such saturation patrols most often produced arrests for minor offenses such as loitering, disorderly conduct, or wearing gang paraphernalia. Typically, the overtaxed Los Angeles
County court system dropped these minor charges almost immediately. Some LAPD officers
were actually dispatched to disrupt truce or peace-making meetings among gangs because they
feared the formulation of megagang alliances.

The Chicago police added their own special touch to gang suppression activities. They arrested
youngsters and then released them in the territories of rival gangs (Klein, 1995). In many cities,
police crackdown campaigns against gangs have led to repeated complaints of police misconduct,
illegal searches, and even planting of illegal evidence. More recently, there have been cities that
utilized "gang injunctions" to fight gangs. The gang injunctions typically involve using municipal
codes or citations to require that alleged gang members not congregate in certain neighborhoods.
The injunctions utilize municipal laws governing "nuisances." What is prohibited is not specific
criminal behavior but rather the mere presence of individuals on streets or other common areas
during specified times of the day. Individuals are somewhat arbitrarily named in these injunctions, although some injunctions named "criminal organizations." Police have used the existence
of gang injunctions to **stop and frisk** young people who may be part of a gang injunction list. If
the alleged offender attempts to resist these police interventions, the legal situation might be escalated to include charges of resisting arrest or impeding police functions. If the young person has
any outstanding warrants from justice agencies, as a potential material witness, or from immigration enforcement officials, these might also be a rationale to take the youth into formal custody.
There have been allegations that police have used gang injunctions to detain youngsters who are
under the authority of child welfare agencies as potential victims of abuse, possible school truants,
curfew violations, or sex trafficking.

There are minimal to zero criteria for proving that someone should be included in these
prohibitions, and almost no evidence must be provided to the courts that oversee these actions.
In a sense, gang injunctions criminalize the constitutionally protected right to assemble. Just
living in a designated area and seeming "suspicious" is enough to get an individual named in a
gang injunction. Once someone gets named in a gang injunction, there is no mechanism to get
off the list. Some cities have published lists of individuals named in gang injunctions, and
this information has been used to deny college admissions, public housing benefits, or employment to those on the lists. There is not solid evidence that gang injunctions have actually reduced
violence or gang-related crimes. Some community residents have criticized these police actions
as, in effect, racial or ethnic profiling. Gang injunctions have been analogized to police practices

in totalitarian countries such as China, Russia, or the Republic of South Africa under apartheid.

A variety of studies showed no public safety benefits or very limited reductions in gang crime. The best research to date on gang injunctions was conducted by Maxson, Hennigan, and Sloane (2005; see also Greene & Pranis, 2007). They found that, at best, gang injunctions produced a very temporary and nonsustained small decline in crime and that there was evidence that the injunctions just moved the crime program to an adjoining area.

Krisberg (2011) and several other well-known gang researchers such as Matthew Klein (1995) and Spergel and Grossman (1996) have noted that aggressive law enforcement tactics that are divorced from community-organizing efforts may actually make the gang problem worse. According to these scholars, police antigang crackdowns reinforce the reputation of gangs and promote greater cohesiveness among gang members. Greater gang cohesion is generally associated with gang members engaging in more violent crimes (Klein, 1995).

Gang suppression efforts have also involved targeted prosecution programs, often in tandem with increased police pressure on gangs. Prosecutors have sought out innovative legal strategies to widen the range of tools that could be used against gangs. For example, the Los Angeles County district attorney asked for a court order declaring gangs to be quasi-corporate entities, thus permitting legal actions against any member based on the activities of the group. One such program required those identified as gang members to participate in a graffiti removal program. Other prosecution efforts tried to declare parks and other public places as gang-free zones. In theory, this permitted the police to arrest gang members simply because they were present at some location, not because they were actually engaging in violations of the law. There have also been efforts to declare juvenile gangs as a new form of organized crime, permitting the use of state and federal racketeering charges to be used to suppress gang activity. More recently, law enforcement officials have been exploring the application of the label "terrorist organization" to some gangs in the hopes that the broad provisions of homeland security legislation that was enacted after the attacks on the World Trade Center in New York on September 11, 2001, could be used to fight urban street gangs. During the 1980s and 1990s, federal crime legislation attempted to authorize the FBI, the Drug Enforcement Administration, the Bureau of Alcohol, Tobacco, Firearms and Explosives, and other federal law enforcement agencies to attack juvenile gangs. There have been several congressional attempts to enact gang suppression or to include dedicated antigang funding through the Office of Juvenile Justice and Delinquency Prevention (OJJDP). The Gang Abatement & Prevention Act introduced by Senator Dianne Feinstein (D–CA) and Representative Adam Schiff (D–CA) sought to expand federal jurisdiction over youth gangs and to stiffen enforcement and punishment for identified gang members. This get-tough approach was countered by Representative Bobby Scott (D–VA) in the Youth Promise Act that focused instead on local authority in framing gang reduction strategies and placed most of its emphasis on delinquency prevention and youth development programs (Vuong & Silva, 2008). Neither bill was passed into law, as the nation became preoccupied with the national financial crisis of 2008. However, to win their political support, some of the provisions of the Feinstein-Schiff bill were incorporated in other federal legislations covering other unrelated topics such as the American Recovery and Reinvestment Act and Affordable Care Act. Repeated attempts to incorporate the American Promise Act into the Juvenile Justice and Delinquency Prevention Act were defeated by more conservative members of congress.

There is little evidence that any of these approaches actually reduced gang activities. However, it is reasonable to assume that young gang members took great pride in how "bad" they were that their activities required the attention of federal law enforcement officers. This fueled their teenage fantasies about being "real Gs" (gangsters). In the youth culture, the word *federal* came to be a compliment—one was truly worthy if the "feds" were after you.

One of the best-known gang prosecution programs was known as Operation Hardcore. This was the prototype for many subsequent, specialized, career-criminal prosecution units. Operation Hardcore involved enhanced training of police and prosecutors in gathering evidence and in the use of expert testimony. Prosecutors were given greatly reduced caseloads so that they could focus their attention on the most notorious gang members. There were provisions for witness protection programs that gave new identities to effective snitches. Operation Hardcore claimed to not engage in plea bargaining and to request very high amounts of bail. The main strategy was vertical prosecution in which one district attorney carried the case forward from beginning to end.

The Mitre Corporation was hired to conduct an evaluation of Operation Hardcore and reported that the program achieved a 95% rate of convictions and longer prison terms for convicted gang members (Dahmann, 1982/1995). The evaluators noted that the intensive staff commitment required by the program and the related high costs meant that Operation Hardcore was applied to just a few gang members. The Mitre Corporation noted that Operation Hardcore achieved a marginal increase in the incapacitation of targeted offenders, but it was far more difficult to prove that a larger deterrent effect was accomplished with other gang members. There was little evidence that Operation Hardcore exerted an impact on the gang problem of Los Angeles that grew throughout the decade of the 1980s (Dahmann, 1982/1995). Operation Hardcore still exists today, but the caseloads of prosecutors assigned to the special unit have grown tremendously, and both conviction rates and average prison terms have gone down.

A variation on the Operation Hardcore model was funded by OJJDP and designed as a national model. Renamed the Habitual, Serious, and Violent Juvenile Offender Program (HSVJOP), this get-tough approach was tested by OJJDP in a number of cities. Cronin, Bourque, Gragg, Mell, and McGrady (1988) conducted a detailed process evaluation of the program, concluding that the targeted prosecution approach was able to overcome the initial resistance of the traditionally treatment-oriented leaders of the juvenile justice system. Further iterations of the model, known as Serious, Habitual Offender Directed Intervention Program and Serious, Habitual Offender Comprehensive Action Program, were funded by OJJDP in several additional locales. These later versions attempted to add treatment components to the core of focused prosecution. These later versions received much less research than the original HSVJOP.

The research found that more experienced prosecutors and more time for case preparation and work with witnesses and victims led to speedier adjudications in some locations. Serious and chronic offenders were more likely to be convicted because of these programs, and there were increases in the numbers of youths who were transferred to criminal courts. It was less clear whether sentences in the juvenile corrections system were toughened. Moreover, OJJDP hoped to implement comprehensive treatment programs that would complement the prosecution approach, but these rehabilitative resources rarely were put in place. Cronin and her colleagues (1988) also noted that the case-screening criteria for HSVJOP were overly broad, producing caseloads that were well beyond the program resources.

The researchers for HSVJOP could not answer a number of crucial public policy concerns. The selective prosecution programs were able to identify offenders with many prior arrests, but it was less clear how well the project's screening criteria would predict future criminal behavior. Cronin and her colleagues (1988) could not demonstrate what case-screening criteria would be optimal in averting future serious youth crime. Chaiken and Chaiken (1987) reached similar conclusions in their study of career criminal prosecutions in Los Angeles, California, and

CASE STUDY: THE GANG BUSTERS

A 16-year-old and his friends were drinking rum. He was involved in carjacking with his friends that resulted in a young mother being shot and killed. The youth did not have a gun and was not the shooter. His role in the crime was minimal. The young person had migrated to the United States as a toddler. He had very little parental supervision and lived in abandoned cars with his younger siblings.

His education was very limited, and he did not finish grade school. He was a very marginal member of a street gang in his neighborhood. He had problems with alcohol use as an adolescent.

The prosecutor filed this case in criminal court and a jury convicted him of murder in the commission of a felony. This meant that the teenager would be expected to die in prison. The judge sentenced the youth to life without the possibility of parole, and he had served almost 25 years in state prison until a new law permitted him to petition for a reconsideration of the sentence. His codefendant pled guilty and served only 6 years before parole. The alleged gang involvement was influential in producing this harsh sentence

During his nearly 25 years in prison, the youth completed his high school degree and 2 years of college courses. He made a sincere commitment to religious practice and expressed remorse for the harm and suffering of the victim's family. His prison behavior was excellent and he was almost free of any disciplinary reports. He was rated as a good and cooperative worker in prison industries. A leading group of civil rights lawyers developed his appeal for resentencing, and despite objections by the local district attorney, the court changed his sentence to allow for parole. The state parole board granted him release at his first hearing before them, and the governor did not reverse this decision.

Middlesex, Massachusetts. The information contained in the case files of the prosecution units was not very useful in discriminating the future between very frequent and less frequent offenders. An analysis of data from the Wolfgang Philadelphia cohort study could not produce a screening tool useful to prosecutors to correctly identify high-rate offenders (Weiner, 1996). The factors that could predict those youths who committed multiple minor crimes were quite different from the factors that isolated those who committed the most serious violent crimes. Weiner (1996) concluded, "There is little reason to believe that warning signs, early or otherwise, can soon be developed that will provide a firm basis for identifying serious or habitual offenders" (p. 1). Further, Delbert Elliott's (1994) research suggests that many offenses committed by juvenile offenders do not result in arrests and that the duration of the careers of high-rate offenders is usually less than one year. Thus, by the time we catch the chronic and serious offenders, their criminal behavior is diminishing in frequency and severity. The sum of these studies suggests that career criminal prosecution programs are of very limited utility in suppressing the future incidence of serious juvenile crime.

Another question that remains unresolved about targeted youth prosecution programs is the impact of the penalties that these programs seek to achieve. A key question is whether traditional juvenile corrections programs or adult-style punishments exert a positive, negative, or neutral impact on criminal careers. If, as some critics have suggested, correctional experiences are **criminogenic**, then the short-term incapacitation effects may be outweighed by longer-term escalation in offending. Getting youthful offenders more quickly into "schools for crime" may be self-defeating. We take up this issue a little later in this chapter.

Not surprisingly, targeted law enforcement and prosecution programs have been very popular among politicians and justice system professionals. However, the sizeable financial investments in these labor-intensive programs have not been matched with serious investments of research dollars to measure their results or refine the state of the art. We know little more today about how to target career offenders than we knew two decades ago. Since crackdowns on gangs and other perceived dangerous offenders are likely to remain popular and to absorb large amounts of funding, it is imperative that these efforts be subjected to far more intensive research efforts than they have been thus far.

The Impact of Juvenile Corrections

Measuring the impact of placing youths in juvenile correctional facilities is no simple task. Juvenile corrections encompasses a very wide range of settings that vary tremendously in terms of security level, size, location, and staffing patterns. The extremes can be represented by 15-bed secure facilities in Massachusetts to institutions of the CYA that each hold more than 200 young people. Juvenile correctional facilities may closely resemble adult prisons and jails, or they may include wagon trains, correctional sailing ships, military-style boot camps, environmental or wilderness programs, group homes, foster homes, chemical dependency residential programs, and independent living arrangements. Most juvenile correctional programs are operated by government agencies, but an increasing number of programs are run by nonprofit organizations and for-profit companies. Most serious and violent juvenile offenders are held in facilities operated by state juvenile correctional agencies, but there has been a growing trend to remove the most dangerous youthful offenders and place them in special institutions that are run by adult departments of correction.

The states differ in their juvenile correctional policies (Krisberg, Litsky, & Schwartz, 1984). Age boundaries between the adult and juvenile correctional systems vary from state to state, and the laws defining these boundaries are changing rapidly (Sickmund & Puzzanchera, 2014). There are states such as Ohio, Pennsylvania, and California in which juvenile facilities are operated by both state and local government agencies. There are other states such as Georgia, Florida, and Louisiana in which all juvenile corrections facilities are managed at the state level. Moreover, states such as Maryland, Utah, Rhode Island, and Massachusetts rely very heavily on nongovernmental programs to run their youth corrections programs. States also vary considerably in how commitment and release decisions are made. In some locales, judges choose the location of incarceration and specify the length of confinement. In other states, institutional placements are determined by the juvenile correctional agencies, sometimes in concert with a juvenile parole board (Dedel, 1998). The nature and extent of aftercare or reentry services that are available to youths upon release is not uniform across jurisdictions.

The general status of juvenile corrections is not good. A national study of juvenile corrections reported that many facilities do not meet even minimal professional standards (Parent et al., 1994). The number of youth in residential facilities has declined dramatically since 1997—declining by 33% between 1997 and 1999. However, many states have closed residential programs or reduced their bed capacities, so that as many as 20% of currently operating youth correctional facilities remain crowded, especially urban juvenile detention centers (Sickmund & Puzzanchera, 2014, p. 204).

Prisons have fared much better in the competition for public funding. There has been very little funding available to juvenile facilities for repairs or replacements of dangerous and dilapidated buildings. The sanitation of youth rooms and day rooms are often deplorable. The rooms are dark and the walls filled with graffiti, the plumbing often does not work properly, and the living areas often smell of spoiled food, human waste, and dirt (Mendel, 2014). Reports of institutional violence and escapes that plagued the early 19th-century houses of refuge continue to the present day (Krisberg, 2013; Krisberg & Breed, 1986; Krisberg, Marchionna, & Hartney 2015). There are numerous credible reports of youth who have been physically and sexually abused by staff and other young people in these facilities. Some elected officials and justice practitioners have even defended these terrible conditions as beneficial to youth in that it deters them from future criminal behavior.

The argument over whether juvenile confinement reduces or increases juvenile offending goes back to the founding of the first juvenile institutions. Advocates of alternatives to incarceration from Charles Loring Brace to Jane Addams to Jerome Miller have asserted that incarceration breeds crime (Krisberg & Austin, 1993). As noted above, defenders of juvenile corrections have argued that confinement, even in terrible conditions, exerts a deterrent effect on offenders (DiIulio, 1995b; Murray & Cox, 1979). Some have claimed that institutional treatment programs can be an effective response to youth crime (Rhine, 1996). This debate has rarely been enlightened by sound empirical data. These extremely poor conditions and public safety results of youth incarceration are explained away as due to inadequate budgets, even though the annual costs of incarcerating youth are generally much more expensive than private schools and colleges (Krisberg, 2016)

Listed below are several studies that document recidivism rates of youth who exit juvenile correctional facilities. There are also studies from Canada and Florida suggesting that incarceration produces worse outcomes for low-level offenders as compared to those who receive home-based sanctions (Aos et al., 2004; Baglivio, 2009; Greenwood, Model, Rydell, & Chiesa, 1996). Gatti, Tremblay, and Vitaro (2009) even suggest that juvenile corrections is "iatrogenic" in the sense that those youth who penetrate deeper into the secure custody system exhibit the worst outcomes in terms of subsequent law violations. Other studies have suggested that released inmates show very high rates of violence toward their children and intimate partners as well as higher levels of homelessness, unemployment, deteriorated personal relationships, and negative self-images (Oliver & Hairston, 2008); Steiner, Garcia, and Mathews (19971995) found that the average stay for minors sent to prison was almost 3 times that of youths charged with very serious offenses who were sent to juvenile facilities. The youths sent to adult prisons had much higher recidivism rates than their juvenile-corrections counterparts (Perkins, 1994). At least one study by Forst, Fagan, and Vivona (1989) suggests that these longer stays are due to the fact that the youngest inmates commit many institutional infractions, thus losing their "good time" release credits and lengthening their period of confinement. These researchers also report that juveniles in adult facilities were less likely to receive treatment services than youths in juvenile corrections. Young people in prison are also less likely to receive education and vocational services. Juveniles in adult facilities more often report being victimized while incarcerated than those in juvenile facilities. Much of this victimization is at the hands of the prison guards (Forst et al., 1989).

A Minnesota study (Podkopacz & Feld, 1996) compared youths who were waived to the adult court to those retained in the juvenile justice system. The transferred youths were much more likely to be incarcerated, and for longer periods, than the juvenile court sample. The transferred youths had a re-arrest rate of 58% compared to 42% for the nontransferred juvenile offenders over the 2-year period that each group was back on the streets. The researchers could not say for sure if the higher recidivism rates for the transfer cases were due to this group's containing more chronic serious offenders, the superior rehabilitative services of the juvenile justice system, or the general failure of severe adult penalties to deter adolescents from committing future crimes.

In another innovative research effort, Fagan (1995) examined the differential methods that New York and New Jersey used to deal with serious and violent juvenile offenders. Due to state laws, juvenile offenders in New York are more likely to be processed in adult courts, whereas comparable New Jersey youngsters would be handled in the juvenile justice system. Fagan looked at nearly 1,200 felony offenders who were ages 15 and 16 and arrested for robbery or burglary in matched New York and New Jersey counties. Interestingly, Fagan found that penalties were both more certain and severe for the New Jersey sample compared with the New York youths. However, he also reported that the New York youths had higher recidivism rates, committed more new crimes, and were crime-free for a shorter time than the New Jersey youths. These results, while intriguing, are difficult to interpret. Were the higher New York recidivism rates due to lesser penalties experienced by the New York youths, the adverse consequences of the adult correctional interventions, or the inability to perfectly match offenders from the two samples?

A far more compelling study by Bishop and her research team produced a more conclusive look at the outcomes of transferred youths in Florida (Bishop, Frazier, Lanza-Kaduce, & Winner, 1996). Nearly 5,500 youths who were transferred and retained in the juvenile justice system were matched on social and demographic factors, as well as a number of legal variables such as the seriousness of the instant charge, the number of current charges, and nature and extent of the youth's prior record. During the follow-up period, the transferred youths had higher failure rates, committed more serious new offenses, and passed less time until the next criminal activity. Bishop and her colleagues conclude that there was little evidence that Florida's aggressive legal policy of trying youngsters in criminal courts had any positive effect on deterring future offending. These researchers concluded that the short-term effects of incapacitating youths for a longer period of time in adult facilities were negated quickly as the youths returned to their home communities and committed many more new offenses than their juvenile court counterparts.

OJJDP funded additional studies of waiver that included a more detailed study in Florida, an analysis of transfer in Utah, and another study in Arizona. Although the results of these studies are still to be published, it seems clear that the findings of the earlier research are confirmed. At best, transfer has no positive deterrent value and may indeed accelerate the criminal conduct of youthful offenders.

SUMMARY

The vast majority of new juvenile justice dollars have gone toward get-tough policies rather than to more treatment services. These policies were designed to get the most serious juvenile offenders off the street and to deter potential offenders through the threat of very harsh penalties. Many of the get-tough programs were aimed at juvenile gangs.

Research has consistently shown that deterrence is best accomplished when sanctions are swift and certain. By contrast, the harsher juvenile justice penalties often are applied in an arbitrary and capricious manner. There is usually a lengthy delay from the time of the offense to the actual beginning of the punishment. Some of these "get tough" approaches, such as gang injunctions, trample on basic legal principles of fairness and due process of law. In most cases, youth of color bear the largest burden for alleged crackdowns on juvenile criminals. Overall, the research on these programs does not support the argument that they are effective. Strong criminal and juvenile justice sanctions possess a natural, intuitive appeal. They permit politicians to talk with macho bravado without being held accountable for the poor results of these programs. Given the enormous fiscal and human investments and human consequences of these approaches, it is scandalous that our research base is so slender. It is as if contemporary policies have adopted a "hear no evil, speak no evil, see no evil" strategy. In particular, policies to increase the use of secure confinement or to place more youths in prison are not supported by scientific evidence but are informed by anecdotes, jingles ("Do the Crime, Do the Time"), and media-popularized fads. Our elected officials go from Tough Love to Scared Straight, from boot camps to chain gangs. It is not surprising that many of our young people have become very cynical about a public policy process that squanders so much money on unproven policies and programs.

REVIEW QUESTIONS

1. What was Operation Hardcore, and what factors made it so ineffective?

2. What are gang injunctions? Are they effective in reducing street violence?

3. What are the principles of the most effective programs to reduce gang behavior?

4. What are the principles of deterrence theory that should guide juvenile justice sanctions?

5. What led to the rise of juvenile boot camps? What were the results of boot camp programs?

6. Does juvenile incarceration suppress or encourage future criminality?

7. Is trying juveniles in criminal courts an effective strategy to reduce juvenile law breaking?

INTERNET RESOURCES

The Campaign for Youth Justice is a national movement dedicated to removing youth from adult corrections facilities and to expand the jurisdiction of the juvenile justice system. They produce a monthly newsletter, policy statements, and original research.
http://www.campaignforyouthjustice.org

The Criminal Justice Legal Foundation is a group of criminal justice officials, victims, and legal advocates who want to increase the punishment and accountability of juvenile offenders.
http://www.cjlf.org

The American Legislative Exchange Council is an association of conservative elected officials that seeks to increase the level and extent of punishment for juvenile offenders. They promote model laws and advocate for getting tough policies in the juvenile and criminal justice system.
http://www.alec.org

The American Civil Liberties Union is the nation's oldest civil rights group. It has chapters across the country that advocate for humane and effective juvenile justice policies. The ACLU has filed numerous lawsuits challenging the abuse of juveniles in the adult and juvenile corrections systems.
http://www.aclu.org

The Campaign for the Fair Sentencing of Youth is a national organization that opposes the sentencing of minors to life without possibility of parole and attempts to discourage laws that put juveniles in adult prisons and jails. This group produces a regular newsletter, promotes research, develops **amicus briefs**, and pushes for more enlightened practices with very young offenders. http://www.fairsentencingofyouth.org

Chapter 10

REDEEMING OUR CHILDREN

EVOLVING STANDARDS OF JUVENILE JUSTICE

For most of recorded history, the standard of justice was based on a harsh regimen of social revenge (Foucault, 1977; Rusche & Kirchheimer, 1939). Stark punishments were designed to instill fear in the hearts and minds of the lower classes. This fear was designed to deter potential lawbreakers and to head off any possibilities of social revolt. Frequent use of the death penalty, especially public executions, was part of the ceremony of justice. Torture of defendants and extreme physical brutality were part of the penal practice of Europe and the New World until well into the 19th century. Thorsten Sellin (2016), in his extraordinary book *Slavery and the Penal System,* demonstrates the similarity in the treatment of slaves and convicts going back to Greek and Roman civilizations. Sellin points out that soldiers captured in wars were quickly converted to slave status and that emerging state authorities used the customs and practices surrounding slavery to establish standards for penal practice. Not surprisingly, the growth of the slave trade to Virginia and other southern colonies came close on the heels of transporting convicts to perform hard labor in America and Australia. The practice of transporting convicts to colonial territories provided European nations with a cheap source of labor and a safety valve to reduce the threats posed by crowded jails (Hughes, 1987).

During this time, there were few distinctions made between children and adults. It was very common for young people to be housed in the same jails and workhouses as adults. English common law set age 7 as a bottom threshold for treating children as criminally culpable. Beyond that principle of culpability, there were few practical differences between children and adults in penal practice until well into the 19th century. Childhood as a concept was not well developed.

Efforts to reform the brutal penal practices of the 18th century were spearheaded primarily by religious groups. The efforts of the Quakers of Pennsylvania to install a new penal philosophy based on religious conceptions of penitence and expiation of sins resulted in the establishment of the first penitentiary in America in the early part of the 19th century

(Barnes, 1926). There were also a few isolated attempts to replace the cruelty of penal practice with more reformative methods that were tied to some Catholic orders in France and Belgium. Coincident with the Quaker reforms in Pennsylvania were the efforts of the religiously motivated reformers who created the first houses of refuge and training schools (Mennel, 1973). Other voices joined in the movement to humanize the punishment system. These advocates relied on enlightenment philosophy to construct a theory of proportionate penalties that they argued were more productive in maintaining social order (Beccaria, 1963; Bentham, 1830). The noted British penal reformer John Howard used both religious arguments and enlightened self-interest to propose massive improvements in the British jail system. Howard noted how diseases such as smallpox, plague, and yellow fever, which were incubating in penal facilities, led to horrendous epidemics that infected free society—an argument that we are hearing again today concerning rates of HIV and hepatitis C among prisoners. The advocates for separating children from adult prisoners suggested that younger offenders were being schooled in criminal behavior by their older inmates. It was also argued that juries would acquit guilty youths because they did not want them housed with adults. Very slowly and inconsistently, in different locales in Europe and America, the regime of penal terror was replaced with other standards of justice.

The most important standard of juvenile justice was the ideal of rehabilitation. The concepts of individualized treatment and personal reform were the cornerstones of the American juvenile court. Interestingly, the ideal of rehabilitation is a relatively recent intellectual development dating back to the Progressive Era at the beginning of the 20th century. Let us recall that the founders of the houses of refuge placed most of their faith in religious instruction, a much regimented daily routine, and hard work. The Elmira Reformatory of the last decade of the 19th century added military drill and physical exercise as part of the reform program. However, it was the advocates for the new juvenile court, such as Julia Lathrop and Jane Addams, who sought to harness the power of the emerging sciences of biology, psychology, sociology, and psychoanalysis to reform wayward youths.

Addams and her associates raised funds to support the work of William Healy to study the cases of thousands of delinquent youths and to create child guidance clinics that were to support the work of juvenile court judges. The core of this approach was in-depth testing and data collection about individual young people. A range of medical, psychological, and social facts were then crafted into a diagnosis and specific treatment plan for each lawbreaker. The goal was to uncover the causes of the delinquent behavior and to alleviate some of those negative forces. Scientific knowledge was to be utilized by the emerging cadre of social workers who would connect with families and youths and provide the needed interventions and services. The rehabilitative ideal rested on a far more optimistic conception of human nature than the punishment model. It assumed that young people were malleable due to their tender ages and developmental potential. It is also important to note that traditional Anglo-American criminal law rests on a crucial assumption that human behavior is guided by free will and that offenders are thus accountable for their actions. The rehabilitative ideal is corrosive of the concept of free will. Under the precepts of the rehabilitation ideal, the offender's free will has been partially or fully impeded by internal and environmental factors. If human behavior can be explained, diagnosed, and treated, then the conventional ideology of free will is called into question. Moreover, our emerging scientific knowledge has made it clear that free will is particularly problematic for young people who have yet to develop fully formed cognitive, emotional, and evaluative abilities.

Unfortunately, the reality of rehabilitation rarely matched its rhetoric. Juvenile courts were never afforded the resources that were actually required to diagnose and treat their young clients. Neither were judges or probation staff given the appropriate training to implement rehabilitative programs. Further, the tragic patterns of abuse in the treatment of incarcerated young people continued despite the incorporation of benign treatment language that masked traditional practices. However, the rehabilitation ideal linked the juvenile court with the broader pursuit of child welfare and contributed many positive system reforms.

America's most famous legal philosopher, Roscoe Pound, went so far as to describe the juvenile court as the greatest step forward in Anglo-American law since the **Magna Carta**. He meant to highlight the unique commitment of the juvenile court to individualized and compassionate responses to youthful offenders. How remarkable to have a legal system that was principally guided by the best interests of the child and not primarily grounded in the nature of the offense! However, the primacy of treating the child might also result in very intrusive interventions by the court in cases in which the legal infractions were fairly minor. Pound was keenly aware that individualized justice, the private nature of juvenile court proceedings, and the lack of emphasis on due process considerations could turn the court into a *star chamber*. Contemporary proponents of a separate judicial system for children continue to struggle with this dilemma (Pound, 1957).

Other early observers of the juvenile court remarked at the internal contradictions that plagued it. For example, sociologist George Hebert Mead (1961) noted that crime control is sought via the "hostile procedures of law and . . . through comprehension of social and psychological conditions" (p. 882). He wryly observed, "The social worker in court is the sentimentalist, and the legalist in the social settlement, in spite of his learned doctrine, is the ignoramus" (p. 882). Mead did not believe that these two alternative approaches ever harmonized. There were also concerns expressed that the broad discretion that was central to the rehabilitative ideal could result in discriminatory practices toward children of color and immigrant children. As we noted earlier, there remain serious concerns that young women are accorded overly harsh and intrusive treatment by the juvenile justice system under the rubric of the concept of rehabilitation.

Beginning in the mid-1960s, the rehabilitative ideal was challenged by legal thinkers, social scientists, and politicians. Court decisions required that the juvenile court be guided by due process and equal protection. In the classic words of U.S. Supreme Court Justice Abe Fortis, "Under our Constitution the condition of being a boy does not justify a kangaroo court" (*In re Gault*, 1967). Juvenile courts and legislatures reluctantly moved to accommodate these constitutional requirements. Simultaneously, social scientists reported a range of research findings that questioned the efficacy of many rehabilitative programs. Robert Martinson (1974) and others raised the specter that few interventions with chronic delinquents actually worked. Another powerful social science paradigm, *labeling theory*, suggested that apparently well-meaning societal interventions actually stigmatized young clients and made them worse. The growing skepticism of the rehabilitative ideal was met with law and order advocates that had always decried the coddling of young criminals by the juvenile court.

The rehabilitative ideal was being attacked on many fronts. On the left, the critics urged policy makers to divert youthful offenders away from the court and to increase the legal assistance available to young defendants. On the right, the call was for harsher punishments and expanded processes to transfer youths to criminal courts and to sentence youth criminals to adult correctional facilities. The emerging standard of justice was a mixture of older notions of deterrence and social revenge, mixed with limited rehabilitative services for "worthy" youngsters. Some called for the juvenile justice system to give out "adult time for adult crime." Treatment services, if offered, were to be backstopped by harsh penalties if young offenders failed to take advantage of these second chances. The media-fueled hysteria over rising juvenile crime rates in the first few years of the 1990s and the political rhetoric demanding tough new measures to control a new generation of supposed superpredators led to the ascendancy of the updated punishment model at all levels of government. It began to look like the juvenile court would soon be abolished. Some of us wondered if the children's court would survive until its centennial celebration in 1999 or if it was already dead.

CONTEMPORARY STANDARDS OF JUVENILE JUSTICE

Just when the juvenile court seemed doomed, two new models of juvenile justice emerged and exerted a profound impact on laws and practice throughout the nation. The first of these approaches was known as Balanced and Restorative Justice (BARJ); the second approach was known as the Comprehensive Strategy for Serious, Violent, and Chronic Offenders (CS). Both models were critiques of the deterrence-focused punishment philosophy that was advocated by more conservative policy makers and practitioners during the Reagan and Bush administrations.

There are several ways in which BARJ and CS could be utilized as complementary standards of juvenile justice. As we will see, BARJ is primarily a way of thinking about the official response to youthful lawbreakers, whereas CS places major emphasis on preventive measures and encompasses a wider range of community institutions. The two models also differ on a number of other dimensions. For example, BARJ is largely a philosophic statement that is not particularly rooted in the research literature, whereas the core of CS is derived from a careful examination of research on the causes and correlates of youth crime, as well as rigorous research on promising prevention and intervention programs. The BARJ model established a central role in the justice process for the victims of juvenile crimes; CS pays scant attention to victims as a key aspect of the reform process. Whereas CS is basically a community-level response to youthful criminality, BARJ concentrates its attention on the individual offender and his or her victim.

SEEKING BALANCE AND RESTORATIVE JUSTICE

Gordon Bazemore and Dennis Maloney (1994) have been the most strenuous advocates of BARJ. Their work began as an attempt to reformulate and strengthen the theory and practice of juvenile probation, but it has enjoyed an excellent reception among legislators, judges, and a broad range of juvenile justice practitioners. Bazemore and Maloney began with the articulation of a set of values and principles that could provide a guiding vision for juvenile justice agencies. This vision was intended to create organizational missions that would determine operational policies and practices.

The big idea underlying BARJ is the notion that crime is injury. BARJ seeks to highlight the hurtful nature of juvenile offending and to define the harm to victims, to the community at large, and to the young lawbreakers. Under the BARJ philosophy, these harms create an obligation to make things right. The traditional paradigm of the juvenile court was guided by considerations of the best interests of the child—there was no particular role in this jurisprudence for discussions about victims' rights or obligations of the child offender to the victims. This is not to say that many juvenile justice agencies paid little attention to victims. Indeed, many of the diversionary programs created by the juvenile justice system made extensive use of monetary compensation to victims or community service as part of their repertoire of sanctions. Moreover, it has become more common for juvenile courts to require monetary payments to victims as a routine part of probation orders, even for youths placed in juvenile correctional facilities.

BARJ created an explicit role for victims and other community representatives in the justice process. It is assumed that the victim's point of view is crucial in fashioning a plan to repair the harm created by the offender. BARJ defines *accountability* as the offender accepting personal responsibility for the harm caused by his or her actions, and the offender is required to actively assist in mitigating that damage. The concept of *restoration* is a core component of BARJ. Restoration implies a process that both repairs the hurt to victims and seeks to rebuild relationships in the community. Advocates of BARJ measure the results of juvenile justice sanctions by how much damage is repaired and how much community relationships are strengthened. Proponents of BARJ are less concerned with the amount of punishment that is exacted. They are also less focused on preventing recidivism, although it is assumed that more effective crime control will emerge from the community involvement and victim restoration.

Figure 10.1 illustrates the three components of BARJ. It is referred to as Balanced and Restorative Justice because it seeks to give equivalent emphasis to offender restoration, public safety, and improving the competency of youthful offenders that enter the juvenile justice system. Operationally, BARJ means the creation of more services for victims, as well as making existing victim services available to all segments of the community. As noted above, BARJ envisions giving victims greater opportunity for their voices to be heard and for them to participate in the juvenile justice system. Increasing community connections requires involving community members

Figure 10.1   The Balanced Approach

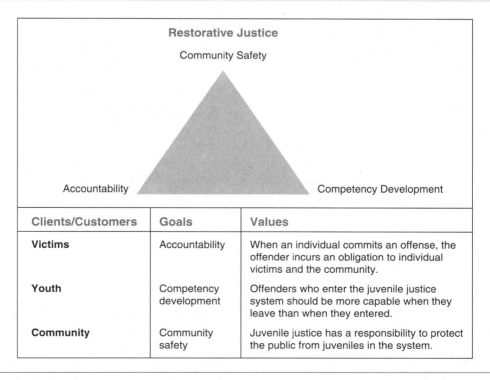

Clients/Customers	Goals	Values
Victims	Accountability	When an individual commits an offense, the offender incurs an obligation to individual victims and the community.
Youth	Competency development	Offenders who enter the juvenile justice system should be more capable when they leave than when they entered.
Community	Community safety	Juvenile justice has a responsibility to protect the public from juveniles in the system.

SOURCE: Office of Juvenile Justice and Delinquency Prevention, Guide For Implementing the Balanced and Restorative Justice Model, (1998, p. 6).

directly in the juvenile justice process. This is a radical departure from traditional notions that keep the community out of juvenile court to protect the confidentiality of young offenders. There is also the implication that ordinary citizens have an important role to play as equal partners with highly professionalized judges, prosecutors, defense attorneys, and probation staff. Offenders are encouraged and given chances to repair the harm that they created. The advocates of BARJ assert that offenders need opportunities to build their skills and abilities so that they can adequately participate in the BARJ process. Core competencies for offenders include educational and vocational advancements and resources to deal with existing family issues, substance abuse problems, or other health issues.

The BARJ movement offers few blueprints for local communities. While many jurisdictions have embraced BARJ in law and agency missions, there is no fully developed implementation to guide replication. BARJ is a value-driven and philosophical approach to establishing a new standard of juvenile justice. Communities are encouraged to use the broad-value framework of BARJ to craft their own approaches.

The open-ended nature of BARJ poses some important concerns. First and foremost, how do you know that BARJ is really being implemented? For example, it is not uncommon for juvenile justice practitioners to assert that they are "already doing BARJ." It remains to be seen how much real innovation will be driven by BARJ. Another concern is the appropriate target population for BARJ approaches: Does it equally apply to juvenile murderers, sex offenders, graffiti artists, drug dealers, and car thieves? BARJ is not rooted in an extensive body of research literature. Indeed, earlier evaluations of victim restitution and community service sanctions for juveniles have produced few positive results in terms of reducing future offending behavior (Krisberg & Austin, 1981). Typically, restorative sanctions have been utilized for minor crimes and as a diversionary option for first-time lawbreakers. There is a large need to fund and complete research on the more recent advances in restorative programming, such as victim–offender reconciliation, community sanctioning panels, and sentencing circles.

Another concern is the definition of *community* that seems so central to BARJ. One can easily imagine how BARJ might be implemented in small, close-knit communities, but the translation of BARJ concepts to highly urbanized areas or the bedroom communities of the suburbs is much more challenging. Moreover, it is not clear that BARJ will be readily accepted in disenfranchised communities that have often been victimized by overly aggressive law enforcement. The social disorganization that plagues many urban areas seems a major hurdle for the implementation of BARJ. Nor is it clear that BARJ will be equally effective in ethnically and culturally differing communities. There seems a real danger that BARJ will advance as the new juvenile justice paradigm for white, middle-class offenders, leaving children of color and poor whites to be relegated to the traditional punishment-oriented incarceration system. Also, there have been concerns expressed that the implementation of BARJ has advanced mostly in the areas of offender accountability and restorative programs. There has been far less attention and funding provided to improving competencies and skills of young offenders.

The future of BARJ is uncertain. The early enthusiasm for its adoption by policy makers and practitioners was supported by the Office of Juvenile Justice and Delinquency Prevention (OJJDP) through conferences, training sessions, and high-quality publications. Expanded juvenile justice funding during the late 1990s provided opportunities for many communities to experiment with BARJ-inspired programs. But hard times in terms of disappearing OJJDP funding and severe fiscal crises at the state and local levels have descended on the nation's juvenile courts. OJJDP is no longer actively encouraging communities to adopt the BARJ model. How well BARJ will fare in this difficult environment remains to be seen.

The core ideals of restorative justice have gained more public awareness and support, especially as the grave problems of "get tough" policies and the problems of mass incarceration have received greater media and political attention. There has been some progress in utilizing the principles of restorative justice in schools to respond to truancy and classroom problems. Some juvenile courts are attempting to employ restorative justice concepts for minor offenders and with probation violators. It is still too early to gauge whether restorative justice principles will gain a greater foothold in the juvenile justice system (Adler School, 2012).

THE COMPREHENSIVE STRATEGY FOR
SERIOUS, VIOLENT, AND CHRONIC JUVENILE OFFENDERS

In the closing days of the administration of George H. W. Bush, two top OJJDP officials, John W. Wilson and James C. Howell, began circulating a policy statement on how to deal with serious, chronic, and violent juvenile offenders. They sought to summarize available research on the topic and to extrapolate principles for improved programming. After significant input from a broad range of researchers and practitioners, they published their conclusions to the field. Wilson and Howell presented a policy statement that was sharply at odds with the views of U.S. Attorney General William Barr. Had President Bush been reelected, this publication would not have been a great career move for them. However, their policy statement was embraced by Attorney General Janet Reno and came to be the official policy view of the Department of Justice. The Comprehensive Strategy on Serious, Violent, and Chronic Juvenile Offenders (CS) was subsequently implemented in nearly 50 communities and emerged as an influential paradigm on juvenile justice (Wilson & Howell, 1993).

As compared with BARJ, which rested on a philosophical foundation, the CS was explicitly linked to research on the causes and correlates of delinquent behavior as well as rigorous program evaluations. The core assumptions of the CS were far more oriented toward prevention than BARJ. Indeed, the CS asserted that prevention was the most cost-effective response to youth crime and that strengthening families was the top priority for prevention programs. The CS also asserted that the core social institutions of the community such as schools, youth-serving organizations, faith-based groups, recreational services, and cultural organizations were key to effective prevention strategies. The CS argued that there was a very small number of serious, violent, and chronic juvenile offenders that needed to be identified and controlled through a range of graduated sanctions. Each community needed to assess its current inventory of prevention and juvenile justice services and to plan to fill gaps in these services. The CS envisioned a continuum of programs and services linking prevention and early intervention programming with juvenile justice responses. The core idea of the CS was to have the right service available to individual youngsters at the appropriate point in their psychosocial developmental stage. Juvenile justice systems needed a mix of immediate sanctions when youths first broke the law, intermediate sanctions for repeat offenders, and secure confinement options for violent youths. Ideal juvenile sanction systems also needed appropriate reentry services for youths who had to be placed out of home. Wilson and Howell urged communities to develop and implement research-based risk and needs assessment tools to guide juvenile justice decision making.

OJJDP issued a request for proposals to take the basic contours of the CS and flesh out the details. The National Council on Crime and Delinquency and Developmental Research and Programs were selected to conduct assessments of the best research-tested prevention and sanctioning programs, to offer communities guidance on planning strategies, and to inventory

CASE STUDY: REDEEMING OUR CHILDREN

An African American 15-year-old girl was arrested for refusing to leave class and go to detention. She allegedly refused to surrender her phone to the teacher. A white school police resource officer attempted to make an arrest, and he was seen on Instagram in cell phone video taken by other students to grab the girl's arm and put his own arm around the student's neck. The video next showed the officer grabbing the girl's hair and throwing her to the ground. She was placed in handcuffs and taken away.

Parents of several students complained to school officials and demanded answers to their questions about the law enforcement officer's behavior. He was suspended pending an investigation, and the case was ultimately referred to the Federal Bureau of Investigation because the local sheriff was concerned about the perception that the review of the incident would not be objective. The school officer was subsequently fired for violating departmental policy in responding to school incidents. The teacher and a school counselor supported the actions of the police officer, but they subsequently received a reprimand by the school board.

existing risk and needs assessment instruments. The resulting product was the *Guide for Implementing the Comprehensive Strategy for Serious, Violent, and Chronic Juvenile Offenders* (Howell, 1995), which was released at a national conference jointly keynoted by Attorney General Janet Reno and Marion Wright Edelman, president of the Children's Defense Fund. The CS guide was later distributed by OJJDP to more than 70,000 juvenile justice professionals and policy makers and was embraced by influential groups such as the National Conference of State Legislatures.

OJJDP and the Jessie Ball duPont Fund provided funding to pilot the implementation of the CS in two Florida communities, Ft. Myers and Jacksonville, and in San Diego, California. Based on very promising results in the pilot sites, OJJDP selected eight states (Florida, Texas, Wisconsin, Maryland, Ohio, Oregon, Iowa, and Rhode Island) to implement the CS in up to six communities in those states. Michigan, Kansas, and Hawaii used their own funds to bring the CS to multiple jurisdictions in those states.

The prevention component of the CS was based on the model of Communities That Care discussed in Chapter 7. The focus was on community-wide planning that identified risk factors that propel youths toward delinquency and protective factors that partially neutralize these risk factors. The graduated sanctions or juvenile justice part of the CS was also based on a risk-focused model that considered factors indicative of serious and repetitive offending. The research suggested that the same protective factors that were effective as prevention resources were also crucial to helping offenders desist from lawbreaking. One of the key insights of the CS was that prevention and intervention strategies should be part of a seamless continuum of youth and family services. The CS implementation attempted to keep the prevention practitioners and the juvenile justice professionals together to conduct joint planning and program development. The CS also relied on empirical data showing that expensive incarceration options were being used for youngsters who could be better managed in less expensive community-based care. Instituting smarter juvenile justice policies and practices could actually free up scarce public dollars to increase investments in prevention services.

The underlying assumption of the CS was that the juvenile justice system should be guided by research and should stress those sanctions shown through research to be most effective. The CS placed a very high priority on public safety as the foremost goal of the juvenile justice system. Research suggested that sanctioning programs that genuinely focused on public protection would rely less on incarceration than current policies that often confined youngsters "for their own protection" or for "treatment." The emphasis on public safety might also go far to reduce the subjective decision making that results in disproportionate minority confinement. Not all adherents of the CS explicitly took this position, however. The logic of the CS required that the first set of justice system decisions had to do with the severity of the current offense (just deserts) and the odds that the youth would recidivate in the future (social defense). After the appropriate level of juvenile justice was determined, based on the above considerations, the CS implied that all youths are entitled to educational, vocational, health care, and counseling services regardless of the gravity of their crimes. Because the criminal court system was ill equipped to provide the appropriate services to young people, this position virtually precluded policies of trying youths as adults or placing them in prisons or jails. In fact, the *Guide for Implementing the Comprehensive Strategy for Serious, Violent, and Chronic Offenders* (Howell, 1995) contained almost no reference to waiver or transfer policies except to say that these practices should be severely limited. The CS offered a range of proven juvenile justice graduated sanctions programs that were effective with the most serious and violent youthful offenders. The CS attempted to sort out the mélange of purposes that were being used by juvenile justice practitioners to explain their actions. These multiple purposes were confusing to workers within the system, to youths and families, and to the public at large. The CS sought to provide a simple rationale for systems of graduated sanctions for juvenile offenders—a rationale that was grounded in solid research.

The CS was met with a very positive reception in virtually all of the communities in which it was implemented. At the local level, there were far fewer shrill voices calling for harsher punishments for young people. Almost all the communities that embraced the CS placed significant attention on expanding prevention and early intervention services. Home-based and family-strengthening services generally received the highest priority in local CS plans. Communities set out to fill key gaps in their local continuum of services. For many localities, it was the first time that they had conceptualized these continua or had applied data-driven comprehensive planning to juvenile crime issues. These communities employed the CS to commence a number of new programs and were able to attract substantial funding for these efforts (Krisberg, Barry, & Sharrock, 2004).

In 2003, OJJDP decided to scrap the CS despite ample evidence that it was helpful to juvenile justice practitioners. OJJDP chose to substitute a more traditional topical approach to providing technical assistance to communities. At this writing, the new model has not been successfully implemented in any location. Many communities are still interested in the CS, but further implementation may be frustrated by current severe budget crises within state and local governments. There have been substantial cutbacks in federal funding for juvenile justice programs. In the wake of the tragic loss of lives on September 11, 2001, Attorney General John Ashcroft directed the Department of Justice to place the highest priority on fighting international and domestic terrorism.

Juvenile justice reform is no longer on the radar screen of federal administration agencies. Congressional leaders who have traditionally supported a strong federal role in juvenile justice seem to be supporting the position of the executive branch. Even as the Obama administration expressed its support for additional federal funding for juvenile justice and prevention programs, federal financial support to states continued to decline. As more conservative politicians dominated the Senate and the House of Representatives, the hope for additional federal financial resources has evaporated. Efforts to reauthorize the JJDPA were introduced as recently as 2015. There have been dedicated lobbying efforts on behalf of the JJDPA by a large number of juvenile justice and advocacy groups. These proposals have even received support by the appropriate Senate committees but have been blocked from a floor vote by several conservative lawmakers. There seems very slim support for the reauthorization of the JJDPA in the House. The Youth Promise Act adopted the CS as its primary approach to local funding, but repeated efforts to enact the Youth Promise Act have failed, and there is little political will to continue the historic federal role of the Juvenile Justice and Delinquency Prevention Act. States and localities have faced enormous fiscal challenges, and support for renewed funding at the state or local level is almost nonexistent. A few states such as North Carolina and Virginia have adopted the principals of the CS model to reduce juvenile incarceration, but fiscal limitations have frustrated efforts to actualize these programs on the community level. Similar to BARJ, the future of the CS as an evolving standard of juvenile justice remains very much in doubt.

Our Children and Other People's Children

Jerome Miller, whose vision led to the closing of the Massachusetts training schools in the early 1970s, often asked his audiences to define the kind of justice system that they would want to encounter if their own child, grandchild, or sibling were in trouble. His assertion was that most people would choose the ideal of the juvenile justice system with its focus on compassion, individualized care, and rehabilitation. Few of us would select an inflexible system of punishments, with little regard for the possibility of personal redemption. Miller asked his listeners to consider that what they want for their own children is what should be available to all children. His point is that only when we delude ourselves into thinking that other people's children deserve a lesser standard of justice can we tolerate the sort of juvenile justice system that we currently have. Elsewhere, I have argued that only racism, class biases, and our unconscious antagonism to troubled young people permit us to consider as rational such draconian juvenile justice polices as trying very young children in criminal courts, sending children to adult jails and prisons, or allowing horrid conditions of confinement to exist in juvenile facilities (Krisberg, 2003).

Much has been made in political debate about the demands of justice. Advocates for dismantling the juvenile justice system have urged us to embrace a single standard of justice for adults and for youths, "Do the Crime, Do the Time." They have advanced arguments that a return to a purely punishment-oriented model will scare young people, deterring them from lawbreaking. Anyone who has taken an introductory psychology course has heard that rewards and reinforcements are a much more powerful method of changing behavior than the calculus of punishments. There is little evidence supporting the claims of the advocates of get-tough policies; moreover, there are important moral reasons why we should not follow this course.

As the horrors of totalitarianism and genocidal hatred descended on Europe, the noted English philosopher Aldous Huxley (1937) wrote that real progress in human societies involved progress in charity.

> In the course of recorded history real progress has been made by fits and starts. Periods of advance in charity have alternated with periods of regression. The eighteenth century was an epoch of real progress. So was most of the nineteenth, in spite of the horrors of industrialism, or rather because of the energetic

Figure 10.2   Overview of Comprehensive Strategy

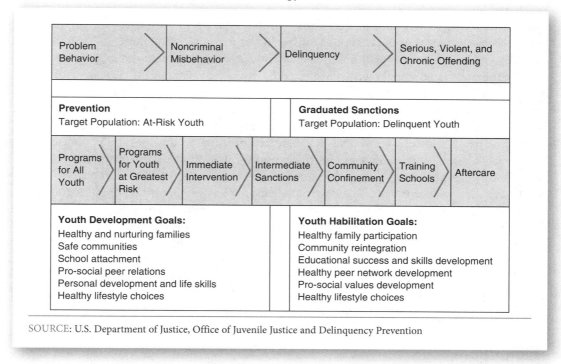

SOURCE: U.S. Department of Justice, Office of Juvenile Justice and Delinquency Prevention

way in which its men of goodwill tried to put a stop to those horrors. The present age is still humanitarian in spots; but where major political issues are concerned it has witnessed a definite regression in charity. (p. 7)

He reminded us that bad means can never be justified by good ends—justice is not an absolute destination that is unrelated to the road traveled to get there.

Huxley urged consideration of a quality of mind that he called being "nonattached." This meant being able to free oneself from angry prejudices, emotional responses to immediate events, and to class prerogatives. For Huxley, the higher ideal meant the application of intelligence to human affairs and the practice of generosity. The nonattached person in Huxley's terms "puts an end to pain; and he puts an end to pain not only in himself, but also, by refraining from malicious and stupid activity, to such pain as he may inflict on others" (Huxley, 1937, p. 6). A generation earlier, founders of the juvenile court such as Jane Addams and Ben Lindsey grasped this wisdom. They sought to establish a new standard of juvenile justice that was redemptive. For these reformers, a redemptive system of justice was the only one that was consistent with democracy.

Can a redemptive standard of justice reemerge in a political climate dominated by fear of outsiders and the genuine sense of personal vulnerability that followed the attacks on the World Trade Center and the Pentagon? There are some hopeful signs in many communities that the original ideal of the juvenile court can be attained, at least partially. We noted that many communities have tried to implement the philosophy of BARJ and that this development may lead to more connectedness among neighbors on behalf of troubled youngsters. In other locales, juvenile court judges have attempted to put into operation ideas of therapeutic justice through the creation of specialized courts that are tailored for drug-involved youths, victims of domestic violence, and young people with serious mental health issues. The Comprehensive Strategy illustrated that communities can coalesce around a common agenda of prevention and improved youth development services.

At present, the topic of juvenile crime is not on media or political agendas. The fear over relatively rare instances of internationally inspired terrorism is getting more attention than the everyday terror experienced by the abused child, by the frightened teen who lacks parental support, by the challenged student who is neglected in school. Some public officials seem lulled into a false sense of security that youth violence rates will continue to fall without our needing to improve the juvenile justice system and the lives of young people. This is probably not a good bet, as the decline in public funding forces major cutbacks in vital educational and social services or as rising unemployment rates put increased pressure on our most vulnerable citizens. Each era creates its own challenges and opportunities. For all of us, especially the politically active and concerned young people whose voices are beginning to be heard, the advancement of Huxley's ideal of charity as the standard of juvenile justice is an important quest. A good place to start is to expose the pernicious fallacy that there are *our* children and *other people's* children. Only when communities embrace all children as our children and work actively for their positive social development can real progress in charity be achieved in the modern age.

In the near term, efforts to reform juvenile justice will likely focus on abolishing the pernicious practices of placing juveniles in adult prisons and in the excessive use of solitary confinement for young people. There will be continued advocacy and litigation to end the extremely abusive practices that continue to dominate juvenile corrections facilities (Krisberg & Vuong, 2009; Mendel, 2011). Courts and legislatures are more open to proposals to end these horrible practices.

SUMMARY

Throughout most of human history, justice has been synonymous with extreme and harsh punishments. These brutal actions were justified based on the theory that draconian penalties would instill fear in the minds of potential lawbreakers. Until the 19th century, there was little distinction in the handling of children by the punishment system. Religious groups, most notably the Quakers, attempted to reduce the severity of the penal system. Others, such as the English reformer John Howard, argued that prisons were the breeding grounds for diseases such as smallpox, yellow fever, and plague that were impacting the free society.

In the 19th century, the founders of the houses of refuge, and later, child advocates such as Jane Addams and Julia Lathrop, furthered the call for enlightened justice by demanding a separate way of dealing with delinquent youths. The ideal of juvenile justice that was established in the first juvenile court laws in America envisioned a model of justice that was individualized, humane, and believed in the possibility of redemption.

By the end of the 20th century, it appeared that the ideal of juvenile justice was in retreat. The dominant political rhetoric favored abolishing the juvenile court and trying children as if they were merely small adults. Just when it looked like the juvenile court was breathing its last sighs, two models emerged to strengthen and reinforce the value of the American juvenile court. The Comprehensive Strategy for Serious, Violent, and Chronic Juvenile Offenders and the paradigm of Balanced and Restorative Justice (BARJ) are two approaches that return to the core principles of Jane Addams and her colleagues while building on the best available research. The Comprehensive Strategy and BARJ have been embraced in hundreds of communities as the appropriate direction for reforming the juvenile justice system. There have been a number of U.S. Supreme Court decisions that have established more humane principles for dealing with youthful offenders, but much work remains to implement these ideas across the nation. The fact that many states are reducing the number of young people growing up behind bars and even closing many secure youth corrections and detention facilities is a hopeful sign.

Although many thinkers such as Roscoe Pound and George Herbert Mead have pointed out how difficult it is to actualize the visions of redemptive justice that were championed by the founders of the juvenile court, the accumulated weight of scientific data supports the efficacy of this approach. Moreover, it is patently clear that most of us would seek a justice system that is founded on core principles of charity and redemption if it were our own children who were in trouble. This, of course, is the key issue. If we recognize the truth that all children are our children, the search for the juvenile justice ideal is our only moral choice.

Translating positive principles of compassionate care of troubled and troublesome young people remains one of the major challenges for the future. In this struggle for justice for youth, we are reminded of the remarkable moral and professional leadership of the great reformer Jerome Miller. Dr. Miller was recruited to reform the Massachusetts youth prisons and had planned to introduce the concepts of **therapeutic communities** for youth in residential placements. His very modest reform efforts to allow youth to wear normal clothing in lieu of jail uniforms to family pictures or books in their rooms and to not cut their hair close to their scalps were met with intense opposition and sabotage by the politically influential corrections workers. There are many reports of Massachusetts Department of Youth Services (DYS) staff physically abusing the youth in their custody.

Miller decided the Massachusetts system had to be radically changed, and he and his colleagues launched a remarkable effort to close down the youth prisons and to move the majority of these youngsters to community-based placements (Miller, 1998), even if Miller found that he could not lay off the state corrections workers unless he could remove the young people from these abusive settings. These events shocked the world of juvenile justice in many states. To withstand the political backlash, Miller launched a masterful campaign to enlist the media and key Massachusetts interest groups to support these major changes. Miller organized press conferences around the Bay State that included former DYS residents. He invited the young people to tell the media "what has been done to you." He enlisted the support of the governor to back his actions.

The continued political opposition of the corrections staff and the opposition to the reforms by many legislators, judges, and prosecutors persuaded Miller that his ability to continue the reforms had diminished. Dr. Miller was subsequently recruited by Pennsylvania and Illinois to continue this "children's crusade" in the child welfare and juvenile justice systems of these states (Krisberg, Marchionna, & Hartney, 2015). The reforms in Massachusetts were largely continued through several subsequent political administrations and inspired juvenile corrections reforms at the federal level and in many states.

For those considering a career in juvenile justice, the lessons of Miller and his colleagues are compelling. First and foremost, the moral imperative is to protect vulnerable young people despite the political risks. Second, effective leaders must not shy away from the media and from political conflict, and they must develop skills of coalition building on behalf of young people. Progressive leaders must take risks to try ideas that are unconventional and to be open to revising plans based on solid facts. Also, it is essential that juvenile justice leaders listen to and give a credible public voice to the young people in their care.

REVIEW QUESTIONS

1. Why do you think Roscoe Pound described the American juvenile court as "the greatest step forward in Anglo-American law since the Magna Carta"?

2. How did the *Gault* case fundamentally change juvenile justice in America?

3. Compare and contrast the Comprehensive Strategy and BARJ.

4. Who was Jane Addams, and what are the core components of her redemptive ideal of juvenile justice?

5. What are lessons to be learned from the closure of the Massachusetts youth prisons?

INTERNET RESOURCES

The Northwestern School of Law Children and Family Justice Center is a major source of information on the evolving standards of justice for youth. They promote research and file class action litigation to defend the rights of juveniles.
http://www.law.northwestern.edu/legalclinic/cfjc

Human Rights Watch is a worldwide organization dedicated to protecting the dignity and rights of vulnerable groups. They produce excellent publications and research reports, and assist in litigation on behalf of the rights of children and their families.
http://www.hrw.org

The Center on the Developing Child at Harvard University is a comprehensive resource library on the latest scientific research on building more promising futures for children and families.
http://www.developingchild.harvard.edu

The North American Family Institute develops very innovative and effective prevention and community-based programs for at-risk youth and their families in many states. They produce a newsletter and excellent program descriptions.
http://www.nafi.com

Southwest Key Programs based in Austin, Texas, operates excellent alternative schools and community-based programs, and conducts economic development projects for impoverished youngsters in several southern states. This organization emphasizes culturally competent programs and employs many clients and their families.
http://www.swkey.org

The Justice Policy Institute is a center for information and research on evidence-based programming and for more humane treatment of juveniles and their families.
http://www.justicepolicy.org

The Center for Children's Law and Policy is an organization that envisions a world in which children in trouble with the law are treated fairly with developmentally appropriate responses that are free of racial and ethnic bias. The center conducts research, promulgates policy positions, and helps inform policies and laws at the state and federal levels.
http://www.cclp.org

Chapter 11

EXPLAINING JUVENILE CRIMINAL BEHAVIOR

Efforts to understand and explain lawbreaking by youth in scientific terms have evolved since the last half of the 19th century. The evolution and popularity of various explanatory paradigms is tied to the emergence of new scientific disciplines and to novel methods of studying human behavior. The process is still moving forward. Prior to the growth of several empirically based studies of delinquency, criminal behavior was explained in religious concepts including demonic possession, the negative influences of secular society, and the inherent evil dwelling within some individuals. Criminal behavior was viewed as a consequence of lacking religious conviction or a basic absence of civilization. Of course, there were obvious contradictions with these theories because their proponents in most European nations did not recognize the value of religions other than Christianity, as well as the legitimacy of some Christian churches. In other nations, a similar viewpoint was held by orthodoxies among Hindus, Muslims, Jews, Buddhists, and others. Moreover, the adherents to this religion-based view failed to acknowledge the horrific crimes that were committed by the religious leadership or against people who were nonbelievers. The rise of the Enlightenment in the 18th century introduced the powerful ideology that humans were inherently good but could be corrupted by adverse social environments, including extreme poverty, alcoholism, and illiteracy.

BIOLOGICAL CAUSES

The last quarter of the 19th century witnessed the beginning of a rudimentary science applied to health issues, intelligence, and medical care. The study of criminal behavior and mental illness was dominated worldwide by the research of Italian army physician Cesare Lombroso who founded the Positivist School of Criminology that substituted empirical studies in lieu of the philosophical speculation that dominated theories of criminality. The Italian physician was especially influential in the United States. Lombroso studied the records of hundreds of prisoners in the hopes of finding common factors among these inmates.

The major focus of Lombroso was on physical abnormalities that could be measured from prison medical records, photographs, and fingerprints. His writings on crime had a far-reaching influence on social policy and the academic approach to crime, as well in the literary world. He concluded that there was a "criminal type" that represented an atavistic or throwback in the evolutionary chain of being (Lombroso, 1911). While emphasizing biological factors, Lombroso wrote about the intertwined influences of social environment and psychology on the causes of criminality.

These pseudoscientific findings were combined with ideas that criminal traits would be passed along through generations and with social Darwinian concepts that human progress could be thwarted for some populations. Lombroso hoped for some empathy for the plight of his "born criminals," but his views favored isolation and confinement of these individuals as well as government efforts to prevent their reproduction through forced sterilization and removal of their offspring to institutions for the criminally insane. Others used Lombroso's ideas to justify savage repression and permanent incarceration for many in the lower classes. Lombroso's ideas were used by many to justify the suppression of racial minorities, colonial domination around the globe, and to argue against basic citizenship for people of color.

Biological theories of delinquency causation still abound, but they lost some of their support among academic criminologists. There continues to be periodic research studies that gain media attention that link youth crime to brain tumors and lesions, a by-product of epilepsy, glandular disorders, mild forms of retardation due to birth defects, chromosomal abnormalities, minimal brain disorders as measured by electroencephalograph patterns, and excessive levels of lead and other chemicals in bodily organs (Rafter, 1997; Rowe, 2001). As with the earliest studies on the biology of crime, most of these data come from very small samples of research subjects that are often elected from incarcerated populations. These students rarely have sufficient research designs that rise to the level of rigorous science. There are a few European studies that try to examine biological factors among identical twins raised in the same family. These studies also have grave scientific limitations.

Other proponents of biological theories of delinquency utilize the flawed method of simply accumulating a large number of inadequate studies, even those in which the initial findings were proven to be falsified (Wilson & Herrnstein, 1985). More recent theorizing in the biology of crime have incorporated a more nuanced perspective that recognizes the interplay of biology, psychology, and the social environment (Raine, 2014). There is also a growing interest in the neuroscience of the brain and how the developing brain of adolescents may be related to their impulsive and risk-taking behavior. Discussions about the implications of adolescent brain functioning have figured in major U.S. Supreme Court cases and are informing legislative debates about the proper ages at which young people can be tried as adults. Some use the early findings of this neuroscience to argue for greater leniency for adolescents and not prosecuting them as criminal offenders, while others may utilize neuroscience research to limit investments in programs that are designed to rehabilitate very young offenders (Zimring, 2016). The interest in neuroscience is gaining traction in juvenile justice policy discussions, but the technology and cost of administering individual brain CAT scans and MRIs will likely limit the practical utility of this biologically focused theory of delinquency causation.

The popularity of biological theories of delinquency is used in public policy discussions about the vast overrepresentation of poor children and youth of color in the justice system. In a nation that espouses the ideology of equal opportunity and fairness, biological factors give comfort to those who do not want to admit to racial and class biases in America. However, the historic connection of biological theories to forced sterilization, incarceration of those with developmental disabilities, denial of education, and the horrors of mass extermination of whole groups of people in Western Europe and many colonies have limited the political currency of biological theories (Hochschild, 1999).

PSYCHOLOGICAL FACTORS

The earliest psychological theories on crime were closely allied with the biological ideas that dominated criminology in the late 19th century, stressing the alleged connections of delinquency to "feeblemindedness," mental defects, and madness. There has been a stream of research attempting, without much success, to link lawbreaking with psychosis, psychopathy, and the structural problems of brain injuries. For some psychologists, such as Hans Eysenck, there was a type of criminal thinking that propelled people toward delinquency, and these psychological patterns were related to low IQ scores (Eysenck, 1997). However, far more influential in juvenile justice are theories that view human nature as evolving and subject to environmental forces. As we noted earlier, the work of William Healy and the Child Guidance Clinic movement were crucial in the founding of the American juvenile court and dominated thinking on treating young offenders throughout most of the 20th century.

Healy was strongly influenced by the writings of Sigmund Freud, and psychoanalytic thinking was central to emerging concepts of treatment and rehabilitation. Later adherents to learning theory and behaviorism had to struggle to gain respectability in the fields of psychology and social work (Rieff, 1959, 1966). The approach to human psychology proposed by Freud and his followers was extremely influential in social work practice and has dominated most court and correctional interventions as a theoretical paradigm, although the actual translation of this theory into practice was, at best, uneven. The Freudian approach to therapy rested in the idea of a highly skilled psychotherapist who spent many hours and even years working with an individual client. It was very costly, and few courts could find the financial resources to deliver these interventions to their primarily indigent clients. Until very recently, government subsidized medical care and most insurance did not cover psychotherapy.

The major assumption of Freudian-influenced psychology was that delinquency emanates from inner conflicts within the child. These inner conflicts may arise due to a number of circumstances, such as lack of parental affectation or excessive parental attention, as well as inconsistent discipline or chaotic parenting practices. Strict Freudians asserted that delinquency could be the result of an unresolved oedipal complex. Others, who considered social factors outside the family such as Erik Erikson, focused on unresolved identity crises caused by extreme poverty, racial or ethnic prejudice, or gender stereotypes that were often beyond the individual's ability to solve (Erikson, 1968). For most youth, the pressure to engage in delinquent or deviant behavior is controlled by external constraints and social learning. This psychodynamic approach postulates that delinquents are inadequately socialized or are not subject to sufficient external pressures on them to conform with general societal expectations. Delinquents are either confused by what ought to be the appropriate cultural norms or have explicitly rejected these values. Chronic delinquents may also become integrated in deviant subcultures such as adolescent gangs that support antisocial attitudes and behaviors. These wayward youngsters may suffer from interpersonal immaturity and lack the social skills to positively interact with others. The proponents of the "inner conflict" theory prescribe individual and group counseling to ameliorate these conflicts. These interventions can include role playing, guided group discussions with other delinquents, or more intensive methods including primal therapy and attack therapy that seek to penetrate surfaced defense mechanisms and force delinquents to confront their innermost feelings.

The psychodynamic approach assumes that the dominant cultural values do not cause law-breaking, although this is a very questionable assumption. Young people are exposed on a daily basis to mass cultural images on television, on the radio, in cinema, and in video games that glorify violence, drug and alcohol abuse, and often equate exploitation of vulnerable people with success in business and in other interpersonal relations. Successful criminals are showcased in the mass media. One might reasonably conclude that the "cultural confusion" comes from the dominant culture rather than a deviant substrata.

Another variation of psychological theories of delinquency rests on the principles of learning theory. These models do not speculate about inner conflicts but rather aim at promoting the desired behavior and suppressing unwanted actions via behavior modification techniques. The focus is on delivering a consistent set of positive and negative reinforcements to troubled youth. This is a very popular method that is utilized in schools and corrections settings where staff equate safety with unquestioning compliance with rules. One encounters a variety of behavior management systems that award points or improved treatment to those youngsters who follow staff instructions or penalize those who express alternative viewpoints. Some of these behavior modification systems may deliver arbitrary and capricious punishments to youth, including extended solitary confinement and physical and chemical restraints. Particularly in correctional facilities, behavior modification regimens can lead to extremely abusive practices (Krisberg, Marchionna, & Hartney, 2015).

There have been several multilayered and eclectic psychological models that are currently very popular in juvenile justice. These ideas are associated with researchers in California, Oregon, Canada, and Ohio. These theories assume that delinquency is a result of criminal attitudes and social factors referred to as *criminogenic needs* that might include a range of family, individual, and social factors. The aim is to identify the most important criminogenic factors for each individual and to help the child or adult reduce the impact of those factors on his or her life. This model also incorporates behavior modification to alter individual attitudes and to permit the individual to be reflective and deliberative in picking the least negative response to a perceived threat or insult (Glick, 2010; Latessa, 2008; Oregon Social Learning Center, 2016). Whereas the prior research on psychodynamic approaches to delinquency has not been encouraging, these later psychodynamic models are gaining more empirical support (Lipsey, Landenberger, & Wilson, 2007).

Individual psychotherapy has fallen out of favor except in older and more affluent communities as many psychologists have embraced cognitive models that can be administered in a group setting, thus cutting down the costs. Within psychiatry, the dominant interventions involve utilizing a range of psychotropic medications to manage symptoms. The use of drug therapies is often paired with cognitive approaches and supportive groups. Moreover, there is growing popularity of self-help groups that engage those with past success in behavior change with those who are still struggling. These groups hold regular meetings and assign mentors to help those who are involved in the process of personal transformation. Several self-help approaches include a very explicit religious orientation such as the 12-step model of Alcoholics Anonymous. These psychological models demand that delinquents publicly acknowledge their own responsibility for bad conduct and often require a formal apology to family members. There is scant research evidence that supports self-help therapies as a long-term solution to personal problems, and there are serious concerns expressed that self-help approaches are not responsive to the needs of particular ethnic groups or to women (Kaskutas, 2009).

A variation of the self-help technique involves working with the family and significant others of the delinquent. It is asserted that the immediate social milieu of the young person matters, and these interventions attempt to build a durable support team among parents, siblings, peers, and other important adult role models. A trained social worker might be assigned to guide the group discussions among members of the support network and to track progress. The most well-known versions of this approach are multisystemic therapy and functional family therapy. These interventions have produced encouraging research results, especially with younger adolescents having severe family conflicts (Barnowski, 2002; Henggeler, Melton, & Smith, 1992).

SOCIOLOGICAL THEORIES OF DELINQUENCY

Sociologists dominated the theories and research on delinquency and deviancy through most of the 20th century. While continuing the focus on explaining the behavior of subcultures, these ideas have been expanded beyond the emphasis on ethnic groups to describe other sorts of subcultures that are made up of gender groups, social class, peer associations, and occupational groups. One observes the reality of a subculture in distinctive language, rituals, style of dress, and the attitudes expressed by members of these groups. Various kinds of subcultural theories have been used to account for an individual's commitment to and participation in delinquent and other behavior that appear to run in opposition to the dominant culture. As societies become less homogeneous and stratified by a range of social groupings, one observes the expansion of diverse definitions of life situations as well as very different ideas of what constitutes appropriate or acceptable conduct.

The sociologist Thorsten Sellin (1938) hypothesized that this differentiation of culture leads to an increase in conflict between groups. Sellin noted that such culture conflict might lead to misunderstandings and fear among adherents to various subcultures. Groups may define the subculture as abnormal and attempt to control or negatively sanction the behavior of those in the subculture. Further, the conflicts in social expectations might lead to internal psychological tensions experienced by those caught between multiple norms defining the "correct" social behavior. Sellin suggested that criminologists devote more attention to the ways in which culture conflicts may produce law violations. This view gained attention as Americans were focusing on the violent criminal behavior of immigrant groups as well as the extreme acts of lynching and violence perpetrated by Southern whites.

One issue that continued to challenge the subculture theories was the growing interpenetration of dominant and subcultural groups. In particular, the powerful influence of the mass media meant that cultural isolation of many groups was breaking down. Moreover, the dominant commercial culture found that it could "monetize" the unique and attractive urban lower-class subculture.

Sellin's formulation of the relationship of culture conflict to crime was substantially expanded through the work of Edwin Sutherland (1947). Sutherland transformed the study of crime and delinquency by boldly asserting that criminal behavior was learned, not inherited through some biological process. Sutherland's conclusions were, in part, based on his studies of professional thieves as well as those who engaged in very dangerous violations of safety rules or financial fraud. He noted that lawbreakers often have to learn the techniques to commit crimes, such as breaking into homes, as well as the ways to dispose of stolen property. Moreover, Sutherland argued that relationships with other delinquents taught the offender the attitudes and values that supported and reinforced breaking the law. The theory of differential association suggested that offenders learned those cultural definitions and values that reduced their fears and self-doubts about criminal behavior. To Sutherland, social learning was the most important element in understanding delinquency and deviance. Individuals learned prosocial and antisocial attitudes from their families, peer groups, and other intimate companions. Sutherland postulated a scheme to calculate the influences of these normative definitions based on the frequency, intensity, and emotional salience of the interactions with other potential offenders. The theory of differential association was no less than a major paradigm shift away from the static psychology and genetic determinism that dominated the study of crime in the first decades of the 20th century. Sutherland's theory was refined and restated in several versions of his classic textbook *Principles of Criminology* that dominated the field for the next half century. The theory of differential association also raised significant doubts as to whether the traditional practice of isolating and confining offenders in penal facilities was iatrogenic.

The emphasis on subcultures was a core element of the emerging study of social deviance at the University of Chicago in the 1920s and 1940s. With the leadership of pioneering sociologists Robert Park, Ernest Burgess, and Roderick McKenzie (1925), a number of sociologists and anthropologists charted a range of distinct social areas within the city of Chicago. Clifford Shaw (1938) concentrated on delinquents, and Thrasher (1927) produced detailed descriptions of gang life in the European immigrant communities. Wirth (1928) explored the social life of the Jewish ghetto, Donald Cressey (1932) produced an excellent profile of taxi-dance hall dancers and commercialized vice, and Zorbaugh (1929) described the ways in which the wealthiest and poorest Chicagoans cooperated in criminal activities. E. Franklin Frazier (1932) and Drake and Cayton (1945) produced rich ethnographies of Chicago's African American communities.

The various research studies produced by University of Chicago sociologists documented city life that was undergoing rapid change due to immigration, demographic changes, urbanization, and economic forces. The pressures of social change contributed to "social disorganization" in which the powerful influences of traditional culture were being eroded. Social disorganization contributed to "cultural confusion" in terms of the appropriate norms that should govern social life and interpersonal relationships. Social disorganization weakened the influence of families, religious institutions, neighborhood associations, and civic groups to assist in the socializing of young people. Increased rates of delinquency was one by-product of this confusion (Shaw & McKay, 1942).

Later sociological theorizing on the etiology of delinquency continued the tradition of examining distinctive subcultures and asserted that the individual learns the skill and values of deviancy from those in the immediate social environment. For example, Albert K. Cohen (1955) extended the idea of subculture beyond ethnic groupings to social classes. While acknowledging that delinquency covers a broad range of behavior, Cohen chose to examine the activities of delinquent gangs in New York City. Cohen suggested that many youthful property offending could be pursued to acquire material possessions. This did not seem to him so hard to explain. Alternatively, the violence exhibited by gangs, the participation in vandalism, and attempts to terrorize others required a deeper explanation for Cohen. Instead, Cohen used the Freudian construct of a "reaction formation" to explain the anger and hostility shown by delinquent gangs. He viewed this anger as a normal psychological defense mechanism by young males who perceived themselves as being rejected and devalued by the dominant social institutions including schools, the labor market, and neighborhood organizations. These antisocial feelings were reinforced and strengthened by the delinquent peer groups that often replaced families and other prosocial individuals as the most salient persons in the alienated young person's life. The policy prescription was clear—attempt to disrupt the gang structures and offer wayward young people alternative ways to achieve positive recognition and emotional support.

For Walter Miller (1958), it was the content of lower-class culture per se that propelled young men toward lawbreaking. He rejected the need to utilize complex ideas about underlying emotional conflicts or psychological disorders to explain delinquency. Looking at the behavior and attitudes of street gangs in Boston, Miller identified what he believed were the tenets or norms characterizing lower-class individuals. These values included the ideas of toughness; smartness, such as the ability to con others; the perpetual pursuit of excitement; and the belief in individual autonomy—one needed to be free of external constraints and not take orders from those higher up in the social class system. Miller argued that lower-class boys felt that "trouble" was a regular part of their lives and that fate or luck dictated how an individual would succeed or fail. Further, the importance of belonging and acceptance of the street-corner group was paramount.

Miller (1958) asserted that following the cultural dictates of the lower-class culture would inevitably lead one to be in conflict with the law. Only luck and personal skill could avert this certain conflict. The attitudes and norms of the lower class were the best explanation of the thinking of delinquents and how they rationalize their conduct. The remedy to this presumed criminogenic lower-class culture was to remove the young person from his immediate social environment, possibly in a correctional school, and to inculcate a whole new culture. The obvious problem with Miller's theory is that persons holding middle-class and upper-class values also engage in serious criminal behavior. Moreover, many of the alleged attributes of lower-class culture have been incorporated in the dominant or standard American value system.

The renowned American sociologist Robert Merton (1938) produced a sophisticated and nuanced view on how conventional values and societal barriers may lead to a range of responses, both deviant and conformist. In Merton's view, the crucial issues involved whether the individual strives to meet the dominant cultural goals and if he or she posseses the ability to meet those objectives through legitimate pursuits. Most people are *conformists* in the sense that they seek conventional ways of achieving success and social repute, and they choose socially acceptable ways of actualizing their goals, such as legitimate work and education advancement. Others, who Merton describes as *retreatists*, neither pursue conventional social values nor act in conventional ways. The prototypes of this group are alcoholic or drug abusers or persons who became enmeshed in religious cult groups. There are also *ritualists* who continue to act in very conforming manners but no longer are clear as to the purpose of their activities. Another group of individuals cling to the dominant goals, and they may use alternative means to achieving those goals. This group, which Merton refers to as *innovators*, can include entrepreneurs, and most important, criminals who engage in a variety of lucrative law violations. Merton also alludes to individuals who seek to challenge traditional norms and also advocate for new methods of achieving these goals—these are the *revolutionaries*. Merton notes that a society that places enormous importance on material possessions and devalues poor people, but also offers few direct pathways to success for the underprivileged, may create severe psychological strains that neutralize the power of social norms to regulate human behavior. He refers to the results as *anomie*—a term denoting the normlessness of people who are subject to radical disruptions in their social lives as a result of wars, severe poverty, natural disasters, or sudden immigration.

Richard Cloward and Lloyd Ohlin (1960), who were both influenced by the Chicago School of Sociology, elaborated the broad outlines of Merton's theory and specified how delinquent subcultures influenced the resulting behavior of persons caught in the functional dilemmas posed by Merton. Cloward and Ohlin emphasized the role of peer groups in promoting and sustaining both prosocial and antisocial norms. Moreover, the peer group offered individuals opportunities to learn and practice legitimate and illegitimate actions. Further, the delinquent peer group offered emotional support for those experiencing anomie, offering potential solutions to the individual's self-doubt and personal turmoil. Cloward urged intervention strategies that both expanded the legal opportunities to succeed and could enlist the peer groups to steer young people toward more legal pursuits.

Sociologists Gresham Sykes and David Matza (1957) expanded the social learning model to examine the ways in which the delinquent subculture provided justifications and rationales for the individual's lawbreaking, thus neutralizing the power of traditional norms. They argued that delinquent youth did not necessarily embrace a distinct subculture that rejected conventional values. Instead, delinquent youth learned a set of justifications or rationales that neutralized their inhibitions against lawbreaking. One powerful technique of neutralization allowed the young person to assert that the bad behavior was not the youth's responsibility and that social factors caused the misconduct. Another powerful rationale was that the harm to the victim was minimal—the victim's life would not be ruined or even substantially injured. Another variation on this theme was that victims deserved to be harmed because they disrespected the offender or engaged in bad behavior themselves. A final technique of neutralization involved an appeal to higher loyalties—I was defending my peers or my family needed more resources to live a decent life. For Sykes and Matza, the delinquent peer group was the learning environment that made it easier for the youth to break the law. These techniques of neutralization did not need to eliminate all guilt or individual anxiety but needed to be sufficient to permit the individual to abridge well-established social norms.

Walter Reckless (1962) opened up a very different line of psychological and sociological inquiry to understand deviant behavior. Reckless chose to examine the forces that led individuals to conform or abide by social norms rather than the subcultural influences that moved individuals toward deviance. Reckless rejected grand theories to explain the etiology of delinquent behavior. Rather, he focused on commonsense ideas that he believed explained the dominance of prosocial attitudes and behavior. He explored external factors that constrained human behavior and internal or psychological factors that reined in delinquent impulses. Implicit in Reckless's containment approach were societal changes that might weaken these pressures to support conformist behavior.

One powerful external constraint was a clear role structure or stratification system that defined the individual's life expectations. Further, this role structure needed to create opportunities for the individual to achieve status within these circumscribed groupings. Reckless also emphasized the need for social cohesion that included joint activities and chances to be together with others in the group. Related to social cohesion was a positive feeling of belonging to the group and ways for the individual to achieve personal satisfaction, even though opportunities to advance in the social structure were closed. Reckless also emphasized the need for "reasonable limits and responsibilities" for group members. Some have interpreted this focus on effective behavior limits as the prime rationale for punishment and proportionate sanctions.

As for internal controls on behavior, Reckless asserted that a positive self-image was important to conforming behavior as well as an individual's capacity to tolerate frustration. Crucial was that the individual held strongly internalized morals and ethics and a very strong and well-developed conscience.

For Reckless, these internal and external constraints buffered the individual from the temptations to break laws and engage in other forms of deviant behavior. He believed that as societies became more complex, mobile, and subject to forces of social change, the internal controls would take on greater significance as group social pressure was attenuated. Reckless proposed that containment theory could be translated into individual assessments that might guide clinical assessments of youth.

The rudimentary ideas of Reckless have been specified and elaborated by many contemporary criminologists. By far, the most influential delinquency theorist was Travis Hirshi (1969). Hirshi noted the prevalence of delinquency in the population and sought to explain the internal and external forces that promoted conformity as opposed to deviance. Hirshi emphasized the key role of societal condemnation and sanctions on inhibiting the natural human tendency to exploit others and to pursue material gain. He also offered a very sophisticated theory of the role played by individual consciousness and emotional support of others in promoting law-abiding conduct. In particular, Hirshi looked at inadequate parenting and incomplete socialization. Social control theory has emerged as a dominant theoretical paradigm in the academy. Later versions placed more emphasis on the influence of positive and prosocial people in the young person's life as well as the importance of providing opportunities for youngsters to bond with others in prosocial activities. From this research tradition has emerged a very popular model that moves away from negative influences and toward an approach to create positive growing-up experiences for all young people (Farrington & Welsh, 2007).

A very different sociological approach to explaining delinquent behavior relied upon the idea that individuals form their identities from experiences and the social reactions to them of those in the social environment. This is a very traditional view of sociology often referred to as *the looking-glass self* (Cooley, 1922; Mead, 1934). In his classic work, Lemert (1951) distinguished between primary versus secondary deviance. He meant to differentiate nonnormative behavior that was the result of normal human impulses from antisocial conduct that was created by the individual being defined as different and limited in their behavior options. This perspective was broadened by Becker (1963) whose perspective suggested that no act is inherently deviant. Becker asserted that behavior became deviant as those with the power to control social definitions chose to *label* the behavior as deviant. Moreover, the person's self-definition is shaped by those significant others who both observed and valued (or devalued) the behavior and the individual. Becker focused on jazz musicians, theatrical performers, and regular drug users who rejected the norms of conventional society and built their own conceptions of what should be defined as "cool" or "smart" (Becker, 1967). Other sociologists applied this perspective to the commercial sex trade, the homeless, the mentally ill, and juvenile gang members (Anderson, 1999; Gove, 1970; Humphreys, 1970; Spradley, 1999).

The labeling paradigm was intended as a powerful rebuke of traditional criminological research that looked to individual failings or deficiencies to explain deviant behavior following Lemert's concept of secondary deviance. Many adherents to **labeling theory** argued that formal government interventions by social workers, police, courts, and corrections agencies would increase the intensity and frequency of deviant conduct rather than suppress or control. Sociologist Edwin Schur (1973) went so far as to advise a policy of "radical non-intervention."

Traditional criminologists attacked the labeling theorists as moral cowards who were unwilling to condemn serious lawbreaking behavior (Akers, 1968; Wilson, 1983). Left-wing scholars decried the labeling approach as understating the significance of class and social oppression in the generation of delinquent and criminal behavior (Liazos, 1974). The labeling theorists did harken back to a vibrant tradition in criminology that emphasized social and class conflicts as central to comprehending law violations and the motives of lawbreakers and law enforcers. This view is known as *conflict theory*.

Conflict theory is closely related to the writings of Karl Marx. The first major statement of Marxist theory in criminology was produced by Willem Bonger (1916). He argued that crime emanated out of the resentment of the lower classes to their exploitation and that poor people committed property crimes to meet their basic economic survival needs. A half century later, the

American criminologist George Vold (1958) elaborated this theory as a way of comprehending the social forces that produced crime. Unlike the subcultural theorists, Vold believed that law-breaking was a manifestation of conflicts among groups who shared a common set of norms or values. Vold asserted that the material interests and needs of different groups led to increasing competition of money, power, and access to social status such as employment and education. The group that was most effective in seizing political power sought to enact laws and to allocate enforcement resources to control the behavior of lesser groups. Vold pointed to the unequal allocation of dollars devoted to suppress theft by blue-collar individuals as compared to the very modest efforts to combat white-collar crime, although upper-class crimes such as stock-market manipulation and financial fraud to the citizenry antitrust violations and environmental crime produced far greater harm than the property crimes committed by poor people.

Another prominent conflict theorist was Austin Turk (1969) who broadened the definition of social strife beyond just economic issues to those that focused on political power, race and ethnicity, gender issues, and differences between the old and the young. Turk also pointed to the exercise of power and influence in daily life that was reflected in religious and educational sectors, as well as the media and even family relationships. Turk believed that power differentials are connected to economic interests but not exclusively so. He also asserted that some forms of social conflict and crime could even be beneficial to the social order by questioning the prevailing norms and social order, and promoting creativity and pressures for social change. For instance, if a significant segment of the populace does not share the dominant norms, efforts to rigorously enforce current laws may be challenged and counterproductive. If there is a social consensus on current norms, law enforcement agencies can operate within substantial community conflict. A good example of this conflict can be seen in the flagrant violations of legal prohibitions on alcohol and some drugs. This argument has also been utilized to encourage police and other justice agencies to reach out to and create positive social bonds in the communities in which they work. Social order is most easily achieved if the least powerful accept the political and economic legitimacy of the status quo and promote the norms of dominant culture throughout all segments of the social structure. This is often referred to as *hegemony*.

The conflict perspective has been extended by some criminologists to argue that serious law violations are produced and justified by criminogenic aspects of the dominant culture. For example, Herman and Julia Schwendinger (1985) show how the mainstream culture as expressed in the quest for material possessions and sexual domination was reflected in peer groups that exploited one another to obtain these culturally defined commodities. Cohen (1974) examined teenage vandalism and violence at sporting events and argued that the rationales for this misconduct were rooted in dominant societal definitions of the appropriate behavior of young men. Krisberg's (1974) ethnography of Philadelphia youthful gang members found that the violence and exploitation exhibited by these groups were explained and justified by cultural themes that are reflected in the dominant culture, not a deviant subgrouping. Others, such as Christie (1975) and Flacks (1971), even suggest that the changing nature of the economy away from agriculture and manufacturing to international trade and advanced technology has created the ambivalent age category of "adolescence" that is rootless and causes young people to seek personal worth in expressions of conspicuous consumption, rage, rebellion, and social deviance. The expanded conflict school in criminology has been further elaborated in the emergence of radical criminology as well as critical criminology (Chambliss & Seidman, 1971; Krisberg, 1975; Taylor, Walton, & Young, 1975).

In recent years, criminology has moved away from debates on general theories of delinquency. The field has been dominated by empirical studies of large groups or cohorts of young people. The most significant study was initiated and subsequently replicated by crimi-nologists Wolfgang, Figlio, and Sellin (1972) and later replicated by their students (Wolfgang, Thornberry, & Figlio, 1987). The focus of the cohort studies was on finding the correlates of the onset and desistence of deviant behavior (Elliott, 1994; Thornberry, Huizinga, & Loeber, 1995). This, in turn, has led to analyses of delinquent careers and the evolution of law viola-tions over the "life course" (Laub & Sampson, 2003). It is interesting that the work of Laub and Sampson relied on a cohort of delinquents from the Cambridge Somerville neighbor-hood near Boston that was first identified in the 1930s.

Adherents to life course or developmental criminology point to significant subgroups within those youth who engage in law violations at some stage in their growing-up period. Within this broad swath of the population (that constitutes most young people, if self-report studies are to be trusted) there are those who begin their delinquent behavior at very young ages as well as those who commence delinquency in later adolescence. There are youngsters who steadily increase their lawbreaking conduct versus those whose deviance is time limited or episodic. There are also suggestions that youth who come in contact with the justice system are more likely to persist in delinquency. This harkens back to the interactionist or labeling perspective in which young people are pushed into groups that reinforce their criminal values and who offer ways to facilitate ongoing or more serious lawbreaking. For the vast majority of youth, they tend to age out of crime as other life pursuits such as marriage, education, and employment compete for their identities and attention. In the main, it is the early starters that are more likely to engage in prolonged and serious delinquency. The one exception is homicide, which seems to be situa-tional and a one-time occurrence. The life course perspective directs our attention to the unique biological, psychological, and social factors that emerge during adolescence and may exert strong pressures of youth to engage in delinquent behavior and other forms of deviance.

Several developmental criminologists have focused on child-rearing patterns, especially poor parenting skills, harsh discipline, and parental antisocial behavior, that increase the odds that a child will start delinquency behavior early and persist in that conduct through adolescence. Moffit (1993) goes as far as theorizing about the neuropsychological vulnerabilities of the child and how these can be negatively affected by environmental factors. Patterson, DeBaryshe, and Ramsey (1989) have also written extensively about the interplay of neuropsychological development and socially toxic growing-up experiences. As noted earlier, there is a resurgence of biologically-based life course studies that attempt to utilize brain scans to illustrate the slow development of the child's brain through the early years and into adolescence. It is argued that brain injuries and the experience of extreme trauma may influence the normal intellectual and emotional development of children. While the implications of this brain research are far reaching in terms of changing how society might deal with the youngest delinquents, the technology to perform this research remains very expensive and not currently applicable to large samples of the youth population. Further, the findings of neurobiology offer a fascinating statistical theory that cannot be applied to individual young people.

It is worth noting that many of the themes and assertions of developmental and life course criminology are reminiscent of the earliest criminological research of Lombroso and Healy. While there is some attention to social factors, this brand of theorizing moves away from the explicit attention to subcultures and peer groups as the primary drivers of youthful misconduct. Moreover, the developmental viewpoint reduces the significance of the social facts that were emphasized by the critical theorists (Krisberg, 1991).

SUMMARY: THE VALUE OF DELINQUENCY THEORY FOR INFORMING SOCIAL POLICY

This chapter offers a range of theoretical speculations on the etiology of delinquent behavior. No single perspective emerges as possessing unequivocal empirical evidence for its assertions. Over time, various theories have gained popular attention in the media and even in the corridors of political power. The rise in influence of a particular theory often has more to do with how well the assumptions of that perspective coincide with the dominant viewpoints of those who have the power to shape public opinion, for example, during the Kennedy and Johnson administrations.

Cloward and Ohlin's (1960) opportunity theory fit well with the ideas of the Great Society and the War on Poverty. Later administrations were far more pessimistic about the possibility of changing the behavior of the lower classes and gravitated to family-centered theories that criticized the negative influences of single parents on their children. For example, during the Reagan and Bush years there was a national movement to decrease the proportion of children that were born out of wedlock and to increase the influence of religion throughout society. During some recent periods of American history, the labeling and interactionist perspective was promoted by those who wanted to reduce the reach of criminal laws and to decriminalize the traditional criminal definitions of conduct including drug abuse, abortion, prostitution, and homosexual behavior. This later view has rarely enjoyed widespread acceptance by political elites, even as public opinion has increased in terms of **decriminalization** ideas, especially drugs such as marijuana.

Criminal laws and the functioning of the criminal justice system remain primarily rooted in ideas favoring the importance of choice and free will in determining who is a deviant. The juvenile justice court has been more open to consideration of family factors on leading to youthful conduct. The juvenile court clings to the rudimentary theory that requiring young people to adhere to strict rules is the best rehabilitative approach. However, most of the sociological theories have received lip service at best. The juvenile court has traditionally depended on the threat of harsh penalties to deter youth from future misbehavior. Harsh treatment of vulnerable young people either via boot camp juvenile programs or incarceration has been the mainstay of the adult and juvenile justice systems. However, recent trends to dramatically reduce the number of young people who are incarcerated have been justified, in part, by renewed appreciation of the labeling perspective. Newer ideas about the neurobiology of adolescence are being discussed as a rationale to reduce the transfer of young people to criminal courts and to end solitary confinement for young people. There is also an emerging awareness of the role of trauma and mental health issues as causes of delinquent behavior.

Theories of delinquency may also help inform programs designed to prevent youthful misconduct before they enter the justice process. Later, we will discuss the current thinking on community-based prevention programming. Not surprisingly, the sociological theories, especially the social control perspective, have gained some foothold in designing prevention efforts.

It must be noted the despite evermore sophisticated and analytic models, developing general theories about youth misconduct are incredibly challenging. If most youths break the law, as reported in several self-report studies, then we need theories that explain the persistence of prosocial conduct, not antisocial behavior. The range of human behavior that is encompassed by most penal codes is much too broad to be explicable by single-factor theories of delinquency. Further, technological developments create new and very serious youth misconduct including cyberbullying, Internet sexual exploitation, terrorist behavior, and computer hacking. Young people may participate in group-facilitated criminal behavior via the Internet and with others whom they have never met. Illegal commodities including weapons, illegal drugs, and other prohibited items can be purchased online. Moreover, gender, racial, and ethnic differences as they operate in dynamic social orders must be included in relevant theories. How well traditional or new theories of delinquency can help us comprehend or control these newer forms of delinquency remains to be seen.

REVIEW QUESTIONS

1. What are the differences between the main biological and psychological theories of delinquency?

2. How do conflict theories differ from the social control perspective?

3. What are the main contributions of the labeling perspective to delinquency theory?

4. What are the life course analyses versus delinquent career research traditions?

5. How does delinquency cohort research differ from studies that examine a 1-day snapshot of juvenile offenders?

6. How do self-report studies versus official data impact theorizing about delinquency?

7. How can delinquency theories be utilized to help shape prevention programming and the treatment of juvenile lawbreakers?

INTERNET RESOURCES

The National Criminal Justice Reference Service is a source of original research on the causes and correlates of youth crime.
http://www.ncjrs.gov

The American Society of Criminology is the leading professional association of researchers in criminology.
https://www.asc41.com

GLOSSARY

Abscond: To fail to appear in court on an appointed day, or to fail to report to one's probation or parole officer or to make them aware of a change in residence.

Accreditation: A formal process to certify competency, credibility, and authority.

Adjudication: A formal court process to determine the guilt or innocence of the defendant, or to accept the child under the jurisdiction of the juvenile court.

Age of Jurisdiction: The age boundaries that qualify a youth for being subject to the authority of the juvenile and criminal courts.

Amicus Brief: Arguments presented by "friends" of the court on points of law that are often matters of broad public interest.

Arraignment: A court proceeding in which the charges are presented and the defendant can plead guilty or innocent.

Bail: A deposit is made to the court for the release of the defendant from custody and to ensure appearance at trial. The amount is set by the court.

Blended Sentencing: Hybrid systems in which either adult or juvenile courts may sentence young people as both adults and juveniles.

Booking: The process by which the jail or detention staff register the charges against a person held for a law violation.

Boot Camp: A program for juvenile or adult offenders characterized by strict discipline, hard physical labor, and community labor.

Capital Punishment: A sentence that requires that the state put the offender to death.

Case Management: A process of identifying clear expectations for behavioral change that should occur within a specific sanctioning option.

Case Management System: A computerized program that assists correctional staff in many aspects of their program including assessments of risk, treatment needs, and supervision strategies. These programs also help track the progress of the youth in meeting the court's objectives.

Caseload: The number of clients that the staff person is assigned to supervise on a regular basis.

Chemical and Mechanical Force: The use of chemical agents such as pepper spray or handcuffs, straightjackets, or shackles to control behavior and to punish individuals.

Civil Rights: Personal liberties afforded to an individual based upon their status as a citizen or resident of a community. These rights are most commonly reflected in the United States or state constitutions or statutes, such as the right to be free from discrimination or the right to not be subject to cruel and unusual punishment.

Classification Systems: A grouping of individuals based upon their risk of future misconduct that permits corrections staff to safely and properly supervise them.

Community Supervision: Various types of noncustodial supervision that allow the offender to live in the community while remaining under the jurisdiction of the court or corrections agency.

Common examples of community supervision include probation or monitoring with the condition that failure to follow these rules might result in custody.

Conditions of Confinement: Overall terms and requirements experienced by persons in custodial settings that are derived from the Eighth Amendment of the U.S. Constitution and include basic rights and humane physical, medical, and psychological care.

Conditions of Probation: The terms and responsibilities that an offender must meet to maintain his or her probation status and to avoid incarceration. These conditions typically include regular contacts with assigned probation officers, avoidance of lawbreaking behavior, not using drugs and alcohol, attending required school, and individual curfews.

Conflict Theory: A vibrant tradition in criminology that emphasized social and class conflicts as central to comprehending law violations and the motives of lawbreakers and law enforcers.

Consent Decree: A legally binding agreement by litigants to resolve a potential lawsuit.

Contempt of Court: A charge of disobeying or disrespecting a valid order by the court that might result in incarceration or a fine.

Conviction: The individual found by a court to be guilty of a criminal offense.

Criminogenic: Tending to give rise to crime rates or an individual's propensity to commit crimes.

Cruel and Unusual: The U.S. Constitution protects individuals from cruel and unusual punishments because they protect the person from undue suffering, pain, or humiliation.

Cultural Competence: The ability to effectively interact with people of different cultures and backgrounds. Requires an awareness of diverse cultures, knowledge of culture and linguistic issues, and an open attitude toward others.

Custody: Detention of a person in a jail, prison, or secure juvenile residential facility by law enforcement officers.

Decriminalization: The process of reducing or eliminating penalties related to specific prohibited conduct.

Dehumanization: Intentional treatment of inmates as less than human by denying them basic services, implementing harsh physical or verbal abuse, and failure to treat people with respect, compassion, or individuality.

Deinstitutionalization: The systematic release of individuals from secure correctional facilities and placing them in the community.

Delinquency: An illegal offense or misdeed committed by a juvenile. Also, habitual misconduct by a youth.

Delinquent Children: Those in violation of criminal codes, statutes, and ordinances.

Dependent Children: Those in need of proper and effective parental care or control but having no parent or guardian to provide such care.

Detention: Holding suspects or defendants as they await further court processing. This usually involves a secure facility and is intended to ensure that the youth attends all of his or her scheduled hearings or does not commit further offenses pending final adjudication.

Deterrence: The use of punishment or threats of punishment to discourage individuals from committing crimes.

Disciplinary Segregation: Placing a youth in a separate cell or room for the purposes of punishment following an incident or rule infraction.

Disparity: Differences in the handling of individuals based on their race, ethnicity, or gender.

Diversion: Programs or policies designed to assist law violators to avoid criminal charges, incarceration, or criminal records.

Due Process Rights: The entitlement or legally binding requirement that the individual will be subject to law enforcement, judicial, and correctional practices that are fair, clear, transparent, and reviewable by objective observers. These rights include the ability to have legal representation, to be heard, to have the proceedings recorded, and not to testify against oneself.

DUI: Driving Under the Influence or Driving While Intoxicated is the crime of operating a motor vehicle with impaired abilities due to alcohol or other regulated drugs. This is usually determined by measuring the level of the prohibited chemicals in the person's blood, or by the inability to perform basic movements.

Eighth Amendment: Prohibits excessive bail, fines, or cruel or unusual punishments.

Evidence-Based Practice: Practices that have been proven to be effective through rigorous and quantitative analysis.

Federalism: The principle of government that defines the relationship between the national government and states. Each government entity has the primary responsibility for certain powers and authority; however, in some instances these may be shared.

Fourteenth Amendment: No state shall make or enforce any laws that abridge the privileges or immunities of citizens of the United States; nor shall any state deprive any person within its jurisdiction of life, liberty, or property without due process and the equal protection of laws.

Gang Injunctions: Laws and municipal codes that prohibit persons who are labeled as gang members to congregate in certain areas.

Gangs: One or more persons who engage in criminal behavior together.

Gender Responsive: Attentiveness, awareness of, and accommodations to the particular needs of women and girls.

Global Positioning System (GPS): Technology used to track the location of persons under correctional supervision, usually in the form of an ankle bracelet.

Graduated Sanctions: A set of correctional responses that increase in severity as the misconduct is repeated or becomes more serious.

Grievance: An official statement of complaint about a wrong done to a person. Facilities must have a policy and procedure for processing and responding to inmate grievances.

Habeas Corpus: A court summons that demands that the custodian present to the prisoner proof of legal authority to detain the person. Habeas corpus is a basic principle of Anglo-American law that attempts to prevent the state from detaining people capriciously or arbitrarily.

House of Refuge: A type of early prison designed for juvenile delinquents; first established in New York City in 1825.

Incapacitation: A way of preventing crimes by removing the offender from the community.

Jail: A local correctional facility that holds persons awaiting their hearings or after conviction. Jail sentences are generally less than one year in duration.

Jim Crow Laws: Laws enacted after the Civil War that enforced racial segregation and helped subjugate racial minorities.

Jury: A group of citizens that assist the court in determining the guilt or innocence of an offender in a criminal trial.

Labeling Theory: A powerful social science paradigm that suggested that apparently well-meaning societal interventions actually stigmatized young clients and made them worse.

Magna Carta: A historic document enacted in 1215 that set out the rights of the English nobility and limited the power of the king.

Malingering: Exaggerating or faking the symptoms of illness to avoid work or school and receive special care.

Neglected Children: Destitute, unable to secure the basic necessities of life, or have unfit homes due to neglect or cruelty.

Net-widening: Applying sanctions to persons who would otherwise be warned or released had these sanctions not existed. This can be an intended consequence of community-based programs.

Neuroscience: A branch of biology that examines the development and the functioning of the human brain. This research is causing a reexamination of traditional ideas about the legal and moral culpability of very young offenders.

Noncriminogenic Needs: These individual characteristics may seem to be related to law violations but have not been proven by research to be associated with crime and delinquency. Examples of these noncriminogenic factors are low self-esteem and depression or being the child of a single parent.

Parens Patriae: An important legal concept that allows the state to assume the parental responsibility for minors who lack effective parental supervision.

Parole: A status of being released from a secure facility after the sentence has been served. The parolee is subject to rules of supervision. In the juvenile justice system, this decision may be made by the sentencing judge, a separate parole board, or the corrections officials themselves.

Penology: The study of punishment and the management of corrections facilities.

Petition: A formal application made in writing to a court to request actions in a specific matter.

Plea Bargain: An agreement between a prosecutor and the defendant whereby the defendant admits to an offense (usually a lesser charge) in return for some concessions by the prosecutor.

Presentence Investigation: A report conducted for the court about an individual who will be sentenced by the court.

Prevention: Programs designed to assist at-risk youngsters from becoming involved in criminal misconduct.

Prison: A correctional facility that is designed to hold those who have been convicted in criminal courts and who will usually serve a sentence of one year or more.

Prison Industrial Complex: The idea that the rapid growth in incarcerations is partly attributable to the lobbying and political influence of private companies or employee unions that benefit financially from the expansion of incarceration.

Privatization: The process of outsourcing government services to private companies.

Probable cause: Sufficient reason based on known facts that a crime has been committed and that the defendant was involved in that offense.

Probation: A court-ordered period of supervision in the community, usually in lieu of confinement.

Prosecutor: The legal entity responsible for presenting a case to the court on behalf of the public against an alleged offender.

Prosocial: Actions and behaviors that are beneficial to the larger society.

Protective Custody: A method of shielding a vulnerable inmate from harm from other prisoners or self-harm, usually via solitary confinement or the use of special housing units.

Racial Profiling: The consideration of race, ethnicity, or national origin by law enforcement officers, based on their belief (often mistaken) that members of these groups are more likely to commit crimes.

Random Sample: A small part of a larger group that is selected to represent the larger entity. Generally, this means that the selection process gives every individual an equal chance of being selected for the study group.

Receiver: A person appointed by the court to oversee an organization to meet its mandated legal responsibilities and financial obligations.

Recidivism: The return to criminal misconduct after been supervised or incarcerated. This is typically measured at various time intervals and may include subsequent arrests, convictions, and return to custody.

Reentry: The process through which a person returns to community living after serving his or her sentence.

Referendum: A form of direct democracy in which the voters may change laws through a ballot measure.

Rehabilitation: Restoring or establishing an offender's ability to contribute positively to society and his or her own well-being through treatment, education, and counseling services.

Residential: Facilities where corrections-involved individuals, especially juveniles, live under 24-hour, seven days a week supervision.

Restitution: A repayment of money or services to the victim or society that may be part of an offender's sentence.

Restorative Justice: A theory of justice that attempts to repair the harm caused to crime victims by cooperative efforts of offenders, victims, and the community.

Retributive Justice: A theory of justice that seeks punishment and exacting the just deserts for a crime.

Revocation: The formal removal of probation or parole status after the court determines that the offender did not meet the conditions of release, usually resulting in incarceration.

Risk Assessment: The process of deciding the level of security or supervision based on an estimate that the individual will engage in future law violations. Often, the corrections officials will utilize a formal risk assessment instrument to inform this decision.

Smart on Crime: A popular phrase that suggests that the best response to crime must consider the severity of the offense, the harm to victims, recidivism reduction, equity, and public safety.

Societal Integration: Breaking down the barriers among different racial, ethnic, and income groups so that the less privileged can move into the mainstream of society.

Solitary Confinement: Special confinement conditions in which the individual is isolated from most human contact. Solitary confinement is used as punishment or to prevent harm to self or others. Research suggests that solitary confinement may create or worsen mental health issues.

Special Master: A representative of the court with specific expertise who assists the court in solving criminal justice and other issues by monitoring and writing reports.

Status Offenses: Behavior that is prohibited only for minors including the consumption of alcohol, smoking, skipping school, chronic arguments with caretakers, breaking curfews, and running away.

Stigma: A perceived or social stain on a person's reputation often unfairly applied due to the prejudices of others.

Stop and Frisk: A program initiated in New York City to temporarily detain and search a person who may have engaged in law violations.

Strip search: The search of a person's body for weapons and contraband that requires the removal of clothing.

Technical Violation: A breach of supervision rules that does not constitute a law violation and would not lead to criminal proceedings. These technical violations are used by parole and probation officers to return the individual to custody.

Therapeutic Communities: A rehabilitative model that emphasizes personal growth and lifestyle changes through self-help, positive reinforcements, and group decision making.

Training School: A residential facility that offers confinement for juvenile offenders. These are usually locked facilities with perimeter security.

Transfer Hearings: A process before a judge to determine if a minor should be prosecuted in the criminal versus juvenile court system. This is sometimes called a "fitness hearing" to determine if the youth could benefit from the services of the juvenile justice system.

Tribal Territories: Also referred to as reservations, these are areas of land that are managed by the federal government and dedicated to Native peoples. These areas are sometimes governed by political entities established by the Native populations.

Waiver: The transfer of juveniles to the criminal court system through laws, court hearings, or prosecutorial charging decisions.

War on Drugs: The policies, laws, and law enforcement practices intended to reduce the trade and use of illegal drugs.

Writ: A court order.

Wrongful Death: A death that results from negligence of criminal justice officials.

Zero Tolerance: A policy, usually in schools, of punishing every act of rule breaking regardless of circumstances or if the actions were accidental.

REFERENCES

Acoca, L. (1998). Outside/inside: The violation of American girls at home, on the streets, and in the juvenile justice system. *Crime and Delinquency, 44*(4), 561–589.

Acoca, L. (1999). Investing in girls: A 21st century strategy. *Juvenile Justice Journal, 6*(1), 3–13.

Acoca, L. (2000). *Educate don't incarcerate: Girls in the Florida and Duval County juvenile justice systems.* San Francisco, CA: National Council on Crime and Delinquency.

Acoca, L., & Austin, J. (1996). *The crisis: Women in prison.* Unpublished manuscript. (Report submitted to the Charles E. Culpeper Foundation, available from the National Council on Crime and Delinquency)

Acoca, L., & Dedel, K. (1998). *No place to hide: Understanding and meeting the needs of girls in the California juvenile justice system.* San Francisco, CA: National Council on Crime and Delinquency.

Addams, J. (1911). *Twenty years at Hull-House.* New York, NY: Macmillan.

Adler, F. (1975). *Sisters in crime.* New York, NY: McGraw-Hill.

Adler School. (2012). *Restorative justice: A primer and exploration of practice across two North American cities.* Chicago, Illinois: Adler School Institute on Public Safety and Social Justice.

Akers, R. (1968). Problems in the sociology of deviance: Social definitions and behavior. *Social Forces, 46,* 455–465.

Alexander, J., Pugh, C., & Parsons, B. (1998). Functional family therapy. In D. S. Elliott (Series Ed.), *Blueprints for violence prevention. Book Three.* Golden, CO: Venture.

Alinsky, S. (1946). *Reveille for radicals.* Chicago, IL: University of Chicago Press.

Altschuler, D., & Armstrong, T. (1984). Intervening with serious juvenile offenders. In R. Mathias, P. Demuro, & R. Allinson (Eds.), *Violent juvenile offenders* (pp. 187–206). San Francisco, CA: National Council on Crime and Delinquency.

American Association of University Women. (1991). *Shortchanging girls, shortchanging America: A nationwide poll that assesses self-esteem, educational experiences, interest in math and science, and career aspirations of girls and boys ages 9–15.* Washington, DC: Author.

American Bar Association & National Bar Association. (2001). *Justice by gender: The lack of appropriate prevention, diversion and treatment alternatives for girls in the justice system.* Washington, DC: Author.

Amnesty International. (1998, November 20). *Betraying the young: Children in the US justice system.* Washington, DC: Author.

Anderson, E. (1999). *The code of the street: Decency, violence and the moral life of the street.* New York, NY: Norton.

Aos, R., Lieb, J., Mayfield, M., Miller, M., & Pennucci, A. (2004). *The costs and benefits or prevention and early intervention programs for youth.* Tacoma, WA: Washington Institute on Public Policy.

Aos, S., Phipps, P., & Korinek, K. (1999). *Research findings on adult corrections' programs: A review.* Olympia: Washington State Institute for Public Policy.

Asian/Pacific Islander Youth Violence Prevention Center. (2001). *Asian/Pacific Islander communities: An agenda for positive action.* Oakland, CA: National Council on Crime and Delinquency.

Asian/Pacific Islander Youth Violence Prevention Center. (2010). *Under the microscope: Asian and Pacific Islander youth in Oakland, needs–issues–solutions.* Oakland, CA: National Council on Crime and Delinquency.

Austin, J., Elms, W., Krisberg, B., & Steel, B. (1991). *Unlocking juvenile corrections: Evaluating the Massachusetts Department of Youth Service.* San Francisco, CA: National Council on Crime and Delinquency.

Austin, J., Krisberg, B., & DeComo, R. (1995). *Juveniles taken into custody: Fiscal year 1993. Statistics report.* San Francisco, CA: National Council on Crime and Delinquency.

Austin, J., Krisberg, B., Joe, K., & Steele, P. (1988). *The impact of juvenile court sanctions.* San Francisco, CA: National Council on Crime and Delinquency.

Baerger, D. R., Lyons, J. S., Quigley, P., & Griffin, E. (2001). Mental health service needs of male and female juvenile detainees. *Journal of the Center for Families, Children, & the Courts* (Judicial Council of California), *3,* 21–29.

Baglivio, M. (2009). The assessment of risk to recidivate among a juvenile offending population. *Journal of Criminal Justice, 37,* 596–697.

Baird, S. C., Ereth, J., & Wagner, D. (1999). *Research-based risk assessment: Adding equity to CPS decision making.* Madison, WI: National Council on Crime and Delinquency, Children's Research Center.

Baird, S. C., Wagner, D., Caskey, R., & Neuenfeldt, D. (1995). *The Michigan Department of Social Services structured decision making system: An evaluation of its impact on child protective services.* Madison, WI: National Council on Crime and Delinquency, Children's Research Center.

Baird, S. C., Wagner, D., Healy, T., & Johnson, K. (1999). Risk assessment in child protective services: Consensus and actuarial model reliability. *Child Welfare, 78*(6), 723–748.

Bakal, Y. (1973). *Closing correctional institutions.* Lexington, MA: Lexington Books.

Barnard, D. E., Booth, C. L., Mitchell, S. K., & Telzrow, R. W. (1988). Newborn nursing models: A test of early intervention to high-risk infants and families. In E. Hibbs (Ed.), *Children and families: Studies in prevention and intervention* (pp. 63–81). Madison, CT: International Universities Press.

Barnes, H. E. (1926). *The repression of crime.* New York, NY: George H. Doran.

Barnowski, B. (2002). *Washington State's implementation of functional family therapy for juvenile offenders: Preliminary findings.* Olympia: Washington Institute for Public Policy.

Bartollas, C., Miller, S., & Dinitz, S. (1976). *Juvenile victimization: The institutional paradox.* New York, NY: Wiley.

Barton, W. H., & Butts, J. A. (1988). *Intensive supervision in Wayne County: An alternative to state commitment for juvenile delinquents. Final report.* Ann Arbor: University of Michigan, Institute for Social Research.

Bazemore, G., & Maloney, D. (1994). Rehabilitating community service: Toward restorative service in a balanced justice system. *Federal Probation, 58,* 24–35.

Beccaria, C. (1963). *On crimes and punishments* (H. Paolucci, Trans.). New York, NY: Botts-Merrill. (Original work published 1764)

Beck, A., Harrison, P., & Guerino, P. (2010). *Sexual victimization of youth in custody.* Washington, DC: U.S. Bureau of Justice Statistics.

Beck, A., Kline, S., & Greenfeld, L. (1988). *Survey of youth in custody.* Washington, DC: U.S. Department of Justice, Bureau of Justice Statistics.

Becker, H. (1963). *The outsiders: Studies in the sociology of deviance.* Glencoe, IL: Free Press.

Becker, H. (1967). Whose side are we on? *Social Problems, 14,* 239–247.

Benson, P. L. (1997). *All kids are our kids: What communities must do to raise caring and responsible children and adolescents.* San Francisco, CA: Jossey-Bass.

Bentham, J. (1830). *The rationale of punishment.* London, England: R. Heward.

Binder, A. (2015, November 10). Alabama 8-year-old charged with murder in toddler's beating. *New York Times.* Retrieved from http://www.nytimes.com/2015/11/11/us/alabama-8-year-old-charged-with-murder-in-toddlers-beating.html

Bishop, D., & Frazier, C. (1992). Gender bias in juvenile justice processing: Implications of the JJDP Act. *Journal of Criminal Law and Criminology, 82,* 1162–1186.

Bishop, D., Frazier, C., Lanza-Kaduce, L., & Winner, L. (1996). The transfer of juveniles to criminal court: Does it make a difference? *Crime and Delinquency, 42*(2), 171–191.

Blumstein, A. (1996). *Youth violence, guns, and illicit drug markets: A summary of a presentation.* Washington, DC: U.S. Department of Justice, Office of Justice Programs, National Institute of Justice.

Bonger, W. (1916). *Criminality and economic conditions.* Vancouver, British Columbia: Political Economy Club.

Borduin, C., Mann, B., Cone, L., Henggeler, S., Fucci, B., Blaske, D., & Williams, R. (1995). Multisystemic treatment of serious juvenile offenders: Long-term prevention of criminality and violence. *Journal of Consulting and Clinical Psychology, 63*, 569–578.

Brace, C. (1872). *The dangerous classes of New York*. New York, NY: Wynkoop & Hallenbeck.

Brager, G., & Purcell, F. (1967). *Community action against poverty*. New Haven, CT: College and University Press.

Bremner, R., Barnard, J., Hareven, T., & Mennel, R. (1970). *Children and youth in America: A documentary history* (Vol. 1). Cambridge, MA: Harvard University Press.

Bridges, G., & Steen, S. (1998). Racial disparities in official assessments of juvenile offenders: Attributional stereotypes as mediating mechanisms. *American Sociological Review, 63*(4), 554–570.

Brown, C. (1966). *Manchild in the promised land*. London, England: Cape.

Brown, L., & Gilligan, C. (1992). *Meeting at the crossroads*. New York, NY: Ballantine.

Brown v. Plata et al., 563 U.S. 1 (2011).

Burns, S. (2011). *The Central Park five: A chronical of a city wilding*. New York, NY: Knopf.

Burns Institute. (2015). *Racial and ethnic disparities in juvenile incarceration and strategies for change*. Oakland, CA. Author.

Butts, J., & Connors-Beatty, D. (1993). The juvenile court's response to violent crime: 1985–1989. Washington, DC: U.S. Department of Justice, Office of Justice Programs, Office of Juvenile Justice and Delinquency Prevention.

Butts, J., & Poe-Yamagata, E. (1993). *Offenders in juvenile court, 1990*. Washington, DC: U.S. Department of Justice, Office of Justice Programs, Office of Juvenile Justice and Delinquency Prevention.

Chaiken, M. R., & Chaiken, J. M. (1987). *Selecting "career criminals" for priority prosecution*. Cambridge, MA: ABT.

Chambliss, W., & Seidman, W. (1971). *Law, order and power*. Reading, MA: Addison-Wesley.

Chavez-Garcia, M. (2012). *States of delinquency: Race and science in the making of California's juvenile justice system*. Berkeley: University of California Press.

Chesney-Lind, M. (1974, July). Juvenile delinquency: The sexualization of female crime. *Psychology Today*, 43–46.

Chesney-Lind, M. (1997). *The female offender: Girls, women, and crime*. Thousand Oaks, CA: Sage.

Chesney-Lind, M., & Shelden, R. (1992). *Girls: Delinquency and juvenile justice*. Pacific Grove, CA: Brooks/Cole.

Chibnall, S. (2001, November). *National evaluation of the Title V Community Prevention Grants Program*. FS 200137. Washington, DC: U.S. Department of Justice, Office of Justice Programs, Office of Juvenile Justice and Delinquency Prevention.

Christie, N. (1975). Youth as a crime-generating phenomena. *New Perspectives on Criminology, 1975*, 1–10.

Cicourel, A. (1968). *The social organization of juvenile justice*. New York, NY: Wiley.

Clark, K., & Myrdal, G. (1965). *Dark ghetto: Dilemmas of social power*. Middletown, CT: Wesleyan University Press.

Cloward, R., & Ohlin, L. (1960). *Delinquency and opportunity*. Glencoe, Illinois: Free Press.

Cloward, R., & Piven, F. (1971). *Regulating the poor*. New York, NY: Pantheon.

Civil Rights of Institutionalized Persons Act, 42 U.S.C. § 1997 *et seq.* (1980).

Coalition for Juvenile Justice. (2000). *Enlarging the healing circle: Ensuring justice for American Indian children*. Washington, DC: Author.

Coates, R., Miller, A., & Ohlin, L. (1978). *Diversity in a youth correctional system: Handling delinquents in Massachusetts*. Cambridge, MA: Ballinger.

Coben, S., & Ratner, L. (Eds.). (1970). *The development of an American culture*. Englewood Cliffs, NJ: Prentice Hall.

Cohen, A. K. (1955). *Delinquent boys*. Glencoe, Illinois: Free Press.

Cohen, S. (1974). Breaking out, smashing up, and the social context of aspiration. *Working Papers in Cultural Studies, 5*, 37–63.

Coles, R. (1967). *Children of crisis: A study of courage and fear*. Boston, MA: Little, Brown.

Commonwealth v. Fisher, 213 Pennsylvania 48 (1905).

Commonwealth v. M'Keagy (1831).

Cooley, C. (1922). *Human nature and the social order*. New York, NY: Scribner.

Crawford, R., Malamud, D., & Dumpson, J. (1970). Working with teenage gangs. In N. Johnson (Ed.), *The sociology of crime and delinquency*. New York, NY: Wiley.

Cressey, P. (1932). *The taxi-dance hall.* Chicago, IL: University of Chicago Press.

Cronin, R., Bourque, B., Gragg, F., Mell, J., & McGrady, A. (1988). *Evaluation of the habitual, serious, and violent juvenile offender program: Executive summary.* Washington, DC: U.S. Department of Justice, Office of Juvenile Justice and Delinquency Prevention.

Dahmann, J. (1982). *Operation Hardcore, a prosecutorial response to violent gang criminality: Interim evaluation report.* Washington, DC: Mitre. (Reprinted from *The modern gang reader* by M. A. Klein, C. L. Maxson, & J. Miller, Eds., 1995, Los Angeles, CA: Roxbury)

Daugherty. C. (2015). *Zero tolerance: How states are complying with PREA's youthful inmate standards.* Washington, DC: Campaign for Youth Justice

Davidson, W. S., Redner, R., Blakely, C. H., Mitchell, C. M., & Emshoff, J. G. (1987). Diversion of juvenile offenders: An experimental comparison. *Journal of Consulting and Clinical Psychology, 55*(1), 68–75.

Davidson, W., Seidman, E., Rappaport, J., Berck, P., Rhodes, W., & Herring, J. (1977). Diversion programs for juvenile offenders. *Social Work Research and Abstracts, 13,* 40–49.

DeComo, R., Tunis, S., Krisberg, B., & Herrera, N. (1993). *Juveniles taken into custody research program: FY 1992 annual report.* Washington, DC: U.S. Department of Justice, Office of Justice Programs, Office of Juvenile Justice and Delinquency Prevention.

Dedel, K. (1998). National profile of the organization of state juvenile corrections systems. *Crime and Delinquency, 44*(4), 507–525.

Developmental Research and Programs. (2000). *Communities that care: Prevention strategies: A research guide to what works.* Seattle, WA: Author.

Devoe, E. (1848). *The refuge system, or prison discipline applied to delinquency* [Sprague Pamphlet Collection]. Cambridge, MA: Harvard Divinity School.

DiIulio, J. (1995a). Crime in America: It's going to get worse. *Reader's Digest,* 55–60.

DiIulio, J. (1995b). Arresting ideas. *Policy Review, 74*(5), 12.

DiIulio, J. (1996). They're coming: Florida's youth crime bomb. *Impact,* 25–27.

Dorfman, L., & Schiraldi, V. (2001). *Off balance: Youth, race & crime in the news.* Washington, DC: Building Blocks for Youth.

Drake, S., & Cayton, H. (1945). *Black metropolis.* New York, NY: Harcourt Brace.

Dryfoos, J. (1990). *Adolescents at risk: Prevalence and prevention.* New York, NY: Oxford University Press.

Du Bois, W. (1903). *The souls of black folk.* New York, NY: Library of America.

Duxbury, E. (1972). *Youth bureaus in California: Progress report #3.* Sacramento: California Youth Authority.

Eigen, J. (1972). Punishing youth homicide offenders in Philadelphia. *Journal of Criminology and Criminal Law, 72,* 1072–1093.

Elikann, P. (1999). *Superpredators: The demonization of our children by the law.* New York, NY: Insight.

Ellingston, J. (1955). *Protecting our children from criminal careers.* Englewood Cliffs, NJ: Prentice Hall.

Elliott, D. (1994). Serious violent offenders: Onset, developmental course, and termination. *Criminology, 32*(1), 1–21.

Elliott, D., & Ageton, S. (1980). Reconciling race and class differences in self-reported and official estimates of delinquency. *American Sociological Review, 45*(1), 95–110.

Elliott, D., Huizinga, D., & Ageton, S. S. (1985). *Explaining delinquency and drug use.* Beverly Hills, CA: Sage.

Emerson, R. (1969). *Judging delinquents: Context and process in juvenile court.* Chicago, IL: Aldine.

Empey, L., & Erickson, M. (1972). *The Provo experiment: Evaluating community control of delinquency.* Lexington, MA: Lexington Books.

Empey, L., & Lubeck, S. (1971). *The Silverlake experiment: Testing delinquency theory and community intervention.* Chicago, IL: Aldine.

English, D. (1997). Current knowledge about CPS decision making. In T. D. Morton & W. Holder (Eds.), *Decision making in children's protective services: Advancing the state of the art.* Atlanta, GA: National Resource Center on Child Maltreatment.

Erikson, E. (1968). *Identity: Youth and crisis.* New York, NY: Norton.

Espiritu, Y. (1992). *Asian American panethnicity: Bridging institutions and identities.* Philadelphia, PA: Temple University Press.

Ex parte Becknell, 51 California 692 (1897).

Ex parte Crouse, 4 Wharton (PA) 9 (1838).

Eysenck, H. (1997). *Dimensions of personality.* New York, NY: Transactions Press.

Fagan, J. (1990). Treatment and reintegration of violent juvenile offenders: Experimental results. *Justice Quarterly, 7*(2), 233–263.

Fagan, J. (1991). *The comparative impacts of juvenile and criminal court sanctions on adolescent felony offenders: Final report.* Washington, DC: U.S. Department of Justice.

Fagan, J. (1995). Separating the men from the boys: The comparative advantage of juvenile versus criminal court sanctions on recidivism among adolescent felony offenders. In J. C. Howell, B. Krisberg, D. Hawkins, & J. Wilson (Eds.), *Serious, violent, and chronic juvenile offenders: A sourcebook* (pp. 238–260). Thousand Oaks, CA: Sage.

Family Services Research Center. (1995). *Multisystemic therapy using home-based services: A clinically effective and cost effective strategy for treating serious clinical problems in youth.* Unpublished manuscript, Department of Psychiatry and Behavioral Sciences, Medical University of South Carolina, Charleston, SC.

Farrell et al. v. Brown, consent decree (2005).

Farrington, D., & Welsh, B. (2007). *Saving children from a life of crime: Early risk factors and effective interventions.* New York, NY: Oxford University Press.

Federal Bureau of Investigation. (1996). *Uniform crime reports: Crime in the United States, 1995.* Washington, DC: Criminal Justice Information Services Division.

Federal Bureau of Investigation. (1997). *Uniform crime reports 1993–1997.* Washington, DC: Criminal Justice Information Services Division.

Federal Bureau of Investigation. (1999). *Crime in the United States, 1998.* Washington, DC: Criminal Justice Information Services Division.

Feld, B. (1977). *Neutralizing inmate violence: Juvenile offenders in institutions.* Cambridge, MA: Ballinger.

Feld, B. (1984). Criminalizing juvenile justice: Rules of procedure for the juvenile court. *Minnesota Law Review, 69*(2), 141–276.

Finckenauer, J., Gavin, P., Hovland, A., & Storvoll, E. (1999). *Scared straight: The panacea phenomenon revisited.* Prospect Heights, IL: Waveland Press.

Flacks, R. (1971). *Youth and social change.* Chicago, IL: Rand McNally.

Forer, L. (1970). *"No one will lissen": How our legal system brutalizes the youthful poor.* New York, NY: John Day.

Forst, M., Fagan, J., & Vivona, T. S. (1989). Youth in prisons and state training schools: Perceptions and consequences of the treatment-custody dichotomy. *Juvenile and Family Court Journal, 81,* 314–347.

Foucault, M. (1977). *Discipline and punish: The birth of the prison.* New York, NY: Pantheon.

Fox, J. A. (1996). *Trends in juvenile violence: A report to the United States attorney general on current and future rates of juvenile offending.* Washington, DC: U.S. Department of Justice, Bureau of Justice Statistics.

Frazier, C. (1991). *Deep end juvenile justice placements or transfer to adult court by direct file?* Tallahassee: Florida Commission on Juvenile Justice.

Frazier, E. F. (1932). *The Negro family in Chicago.* Chicago: University of Chicago Press.

Freitag, R., & Wordes, M. (2001). Improved decision making in child maltreatment cases. *Journal of the Center for Families, Children & the Courts, 3,* 75–85.

G.F. et al. v. Contra Costa County, consent decree (2015).

Gatti, U., Tremblay, R., & Vitaro, F. 2009. Iatrogenic effect of juvenile justice. *Journal of Child Psychology and Psychiatry, 50,* 991–998.

Gendreau, P., Little, T., & Goggin, C. (1996). A meta-analysis of the predictors of adult offender recidivism: What works. *Criminology, 34*(4), 575–607.

Gendreau, P., & Ross, R. (1987). Revivification of rehabilitation: Evidence from the 1980s. *Justice Quarterly, 4,* 349–407.

Glick, B. (2010). *Aggression replacement training: A comprehensive intervention for aggressive youth.* Champaign, IL: Research Press

Glueck, S., & Glueck, E. (1934). *One thousand juvenile delinquents: Their treatment by court and clinic.* Cambridge, MA: Harvard University Press.

Goodstein, L., & Sontheimer, H. (1987). *A study of the impact of ten Pennsylvania residential placements on juvenile recidivism.* Shippensburg, PA: Center for Juvenile Justice Training and Research.

Gossett, T. (1963). *Race: The history of an idea in America.* Dallas, TX: Southern Methodist University Press.

Gottfredson, D., & Barton, W. (1992). *Deinstitutionalization of juvenile offenders.* College Park: University of Maryland.

Gove, W. (1970). Societal reactions as an explanation of mental illness. *American Sociological Review, 35,* 873–884.

Gray, M. (2013, January 8). Q&A: The wrongly convicted Central Park Five on their documentary, delayed justice and why they're not bitter. *Time.* Retrieved from http://entertainment.time.com/2013/01/08/qa-the-central-park-five-on/

Greene, J., & Pranis, K. (2007). *Gang wars: The failure of enforcement tactics and the need for effective public safety strategies.* Washington, DC: Justice Policy Institute.

Greene, Peters, & Associates. (1998). *Guiding principles for promising female programming: An inventory of best practices.* Washington, DC: U.S. Department of Justice, Office of Justice Programs, Office of Juvenile Justice and Delinquency Prevention.

Greenwood, P., Model, K., Rydell, C., & Chiesa, J. (1996). *Diverting children from a life of crime: Measuring costs and benefits* (Rev. ed.). Santa Monica, CA: Rand.

Greenwood, P., Rydell, C., Abrahamse, A., Caulkins, J., Chiesa, J., Model, K., & Klein, S. (1994). *Three strikes and you're out: Estimated benefits and costs of California's new mandatory-sentencing law.* Santa Monica, CA: Rand.

Greenwood, P., & Turner, S. (1993). Evaluation of the Paint Creek Youth Center: A residential program for serious delinquents. *Criminology, 31*(2), 263–279.

Greenwood, P., & Zimring, F. (1985). *One more chance: The pursuit of promising intervention strategies for chronic juvenile offenders.* Santa Monica, CA: Rand.

Haapanen, R. (1988). *Selective incapacitation and the serious offender: A longitudinal study of criminal career patterns.* Sacramento: California Department of the Youth Authority, Program Research and Review Division.

Hahlweg, D., Markman, H., Thurmaier, F., Engl, J., & Eckert, V. (1998). Prevention of marital distress: Results of a German prospective longitudinal study. *Journal of Family Psychology, 12*(4), 543.

Haley, A. (1965). *The autobiography of Malcolm X, with assistance from Alex Haley.* New York, NY: Grove Press.

Haney, C. (2002, January). *The psychological impact of incarceration: Implications for post-prison adjustment.* Paper presented at the From Prisons to Home Conference, National Institutes of Health, Bethesda, MD.

Hammond, W., & Yung, B. (1993). *Evaluation and activity report: Positive adolescents choices training program (PACT).* Dayton, OH: Wright State University, School of Professional Psychology.

Hamparian, D. (1982). *Major issues in juvenile justice information and training. Youth in adult courts: Between two worlds.* Washington, DC: U.S. Government Printing Office.

Hamparian, D., & Leiber, M. (1997). *Disproportionate confinement of minority juveniles in secure facilities: 1996 national report.* Champaign, IL: Community Research Associates.

Handlin, O. (1959). *The newcomers.* New York, NY: Doubleday.

Hartney, C., Wordes, M., & Krisberg, B. (2002). *Health care for our troubled youth: Provision of services in the foster care and juvenile justice systems of California.* Commissioned by the California Endowment. Oakland, CA: National Council on Crime and Delinquency.

Hawes, J. (1971). *Children in urban society: Juvenile delinquency in nineteenth century America.* New York, NY: Oxford University Press.

Hawkins, J. D., & Catalano, R. F. (1992). *Communities that care: Action for drug abuse prevention.* San Francisco, CA: Jossey-Bass.

Hawkins, J. D., & Weis, J. G. (1985). The social development model: An integrated approach to delinquency prevention. *Journal of Primary Prevention, 6*(2), 73–97.

Healy, W. (1915). *The individual delinquent: A textbook of diagnosis and prognosis for all concerned in understanding offenders.* Boston, MA: Little, Brown.

Healy, W. (1917). *Mental conflicts and misconduct.* Boston, MA: Little, Brown.

Healy, W., Bronner, A., & Shimberg, M. (1935). The close of another chapter in criminology. *Mental Hygiene, 19,* 208–222.

Henggeler, S. W., Melton, G. B., & Smith, L. A. (1992). Family preservation using multisystem therapy: An effective alternative to incarcerating serious juvenile offenders. *Journal of Consulting and Clinical Psychology, 60,* 953–961.

Henry, B., Moffitt, T. E., Robins, L., Earls, F., & Silva, P. A. (1993). Early family predictors of child and adolescent antisocial behavior: Who are the mothers of delinquents? *Criminal Behavior & Mental Health, 3,* 97–118.

Herrnstein, R. J., & Murray, C. (1994). *The bell curve: Intelligence and class structure in American life.* New York, NY: Free Press.

Higham, J. (1971). *Strangers in the land.* New York, NY: Atheneum.

Hirshi, T. (1969). *The causes of delinquency.* Berkeley: University of California Press.

Hochschild, A. (1999). *King Leopold's ghost: A story of greed, terror and heroism in colonial Africa.* New York, NY: Houghton Mifflin.

Horachek, H. J., Ramey, C. T., Campbell, F. A., Hoffman, K. P., & Fletcher, R. H. (1987). Predicting school failure and assessing early intervention with high risk children. *Journal of the American Academy of Child and Adolescent Psychiatry, 26*(5), 758–763.

Howell, J. C. (Ed.). (1995). *Guide for implementing the comprehensive strategy for serious, violent, and chronic juvenile offenders.* Washington, DC: Office of Justice Programs, Office of Juvenile Justice and Delinquency Prevention.

Howell, J. C. (1997). *Juvenile justice and youth violence.* Thousand Oaks, CA: Sage.

Howell, J. C. (1998). *Youth gangs: An overview.* Washington, DC: U.S. Department of Justice, Office of Justice Programs, Office of Juvenile Justice and Delinquency Prevention.

Howell, J. C. (2003). *Preventing & reducing juvenile delinquency: A comprehensive framework.* Thousand Oaks, CA: Sage.

Howell, J. C., Krisberg, B., Hawkins, J. D., & Wilson, J. J. (Eds.). (1995). *Serious, violent, & chronic juvenile offenders: A sourcebook.* Thousand Oaks, CA: Sage.

Hoyt, S., & Scherer, D. G. (1998). Female juvenile delinquency: Misunderstood by the juvenile justice system, neglected by social science. *Law and Human Behavior, 22*(1), 81–107.

Hoytt, E. H., Schiraldi, V., Smith, B. V., & Ziedenberg, J. (2001). *Reducing racial disparities in juvenile detention.* Baltimore, MD: Annie E. Casey Foundation.

Hsia, H. M. (2001, May). *An overview of the Title V community prevention grants program.* Washington, DC: U.S. Department of Justice, Office of Justice Programs, Office of Juvenile Justice and Delinquency Prevention.

Hubner, J., & Wolfson, J. (1996). *Somebody else's children: The courts, the kids, and the struggle to save America's troubled families.* New York, NY: Crown.

Hughes, R. (1987). *The fatal shore.* New York, NY: Knopf.

Huizinga, D., & Elliott, D. S. (1987). Juvenile offenders: Prevalence, offender incidence, and arrest rates by race. *Crime and Delinquency, 33*(2), 206–223.

Humphreys, L. (1970). *Tearoom trade.* Chicago, IL: Aldine.

Huxley, A. (1937). *Ends and means.* New York, NY: Harper & Brothers.

In re Gault, 387 U.S. 1 (1967).

In re Winship, 397 U.S. 358 (1970).

Jackson v. Hobbs, No. 10-9647 (Ark. S. Ct. 2012).

J.B.D. v. North Carolina, 131 S. Ct. 2394 (2011).

Johnson, L. D., & Bechman, J. G. (1981). *Highlights from student drug use in America 1975–1980.* Washington, DC: U.S. National Institute on Drug Abuse, U.S. Government Printing Office.

Jones, M. A., & Krisberg, B. (1994). *Images and reality: Juvenile crime, youth violence, and public policy.* San Francisco, CA: National Council on Crime and Delinquency.

Jordan, W. (1974). *The white man's burden.* New York, NY: Oxford University Press.

Justice Policy Institute. (2011). *Arresting education: The case against police in schools.* Washington, DC: Justice Policy Institute.

Juszkiewicz, J. (2000). *Youth crime/adult time: Is justice served?* Washington, DC: Building Blocks for Youth.

Kaskutas, L. (2009). Alcoholics Anonymous effectiveness: Faith meets science. *Journal of Addictive Diseases, 28,*145–157.

Katz, J. (1988). *Seductions of crime: Moral and sensual attractions in doing evil.* New York, NY: Basic Books.

Kelley, B., Thornberry, T., & Smith, C. (1997). *In the wake of childhood maltreatment.* Washington, DC: U.S. Department of Justice, Office of Justice Programs, Office of Juvenile Justice and Delinquency Prevention.

Kent v. United States, 383 U.S. 541 (1966).

Kittrie, N. (1971). *The right to be different: Deviance and enforced therapy.* Baltimore, MD: Johns Hopkins Press.

Klein, M. (1969, July). Gang cohesiveness, delinquency, and a street-work program. *Journal of Research on Crime and Delinquency,* 143.

Klein, M. W. (1995). *The American street gang: Its nature, prevalence, and control.* New York, NY: Oxford University Press.

Kobrin, S. (1970). The Chicago area project: A twenty-five year assessment. In N. Johnson (Ed.), *The sociology of crime and delinquency.* New York, NY: Wiley.

Kolbo, J., Blakely, E., & Engleman, D. (1996). Children who witness domestic violence: A review of empirical literature. *Journal of Interpersonal Violence, 11*(2), 281–293.

Krisberg, B. (1971). *Urban leadership training: An ethnographic study of 22 gang leaders* (Unpublished doctoral dissertation). University of Pennsylvania, Philadelphia.

Krisberg, B. (1974). Gang youth and hustling: The psychology of survival. *Issues in Criminology, 9,* 115–129.

Krisberg, B. (1975). *Crime and privilege.* Englewood Cliffs, NJ: Prentice Hall.

Krisberg, B. (1981). *National evaluation of prevention: Final report.* San Francisco, CA: National Council on Crime and Delinquency.

Krisberg, B. (1991). Are you now or have you ever been a sociologist? *Journal of Criminology and Criminal Law, 82,* 141–155.

Krisberg, B. (1992). *Excellence in adolescent care: The Thomas O'Farrell Youth Center (TOYC), Marriottsville, MD.* Oakland, CA: National Council on Crime and Delinquency.

Krisberg, B. (1997). *The impact of the justice system on serious, violent, and chronic juvenile offenders.* San Francisco, CA: National Council on Crime and Delinquency.

Krisberg, B. (2003). The end of the juvenile court: Prospects for our children. In D. Hawkins, S. Meyers, & R. Stone (Eds.), *Crime control and social justice: The delicate balance.* Westport, CT: Greenwood Press.

Krisberg, B. (2011). *Expert declaration opposing a preliminary injunction in People of California ex rel Russo v. Fruitvale Nortenos et al.* Oakland, CA: Alameda Superior Court.

Krisberg, B. (2013, December 31). *Let's end the lawlessness in the criminal justice system.* Retrieved from http://thecrimereport.org/2013/12/31/2013-12-lets-end-lawlessness-in-the-justice-system/

Krisberg, B. (2016). Reforming juvenile facilities. In P. Sturmey (Ed.), *The handbook of violence prevention.* New York, NY: Wiley.

Krisberg, B., & Austin, J. (1981). Wider, stronger and different nets: The dialectics of criminal justice reform. *Journal of Research on Crime and Delinquency, 18,* 165–196.

Krisberg, B., & Austin, J. (1993). *Reinventing juvenile justice.* Thousand Oaks, CA: Sage.

Krisberg, B., Austin, J., Joe, K., & Steel, P. (1988). *A court that works: The impact of juvenile court sanction.* San Francisco, CA: National Council on Crime and Delinquency.

Krisberg, B., Austin, J., & Steele, P. (1991). *Unlocking juvenile corrections.* San Francisco, CA: National Council on Crime and Delinquency.

Krisberg, B., Bakal, Y., DeMuro, P., & Schiraldi, V. (2015, August 18). *Remembering Jerome Miller: Juvenile justice revolutionary.* Retrieved from http://thecrimereport.org/2015/08/18/2015-08-remembering-jerome-miller-a-juvenile-justice-revolut/

Krisberg, B., Barry, G., & Sharrock, E. (2004). *Reforming juvenile justice through comprehensive community planning.* Oakland, CA: National Council on Crime and Delinquency.

Krisberg, B., & Breed, A. (1986, December). Juvenile corrections: Is there a future? *Corrections Today, 48*(8), 14–20.

Krisberg, B., Currie, E., Onek, D., & Wiebush, R. (1995). Graduated sanctions for serious, violent, and chronic juvenile offenders. In J. C. Howell, B. Krisberg, J. D. Hawkins, & J. J. Wilson (Eds.), *Serious, violent, & chronic juvenile offenders: A sourcebook* (pp. 142–170). Thousand Oaks, CA: Sage.

Krisberg, B., DeComo, R., Wordes, M., & Del Rosario, D. (1995). *Juveniles taken into custody research program: FY 1995 annual report.* San Francisco, CA: National Council on Crime and Delinquency.

Krisberg, B., Litsky, P., & Schwartz, I. (1984). Youth in confinement: Justice by geography. *Journal of Research in Crime and Delinquency, 21*(2), 153–181.

Krisberg, B., Marchionna, S., & Hartney, C. (2015). *American corrections: Concepts and controversies.* Thousand Oaks, CA: Sage.

Krisberg, B., Noya, M., Jones, M., & Wallen, J. (2001). *Juvenile detention alternatives initiative: Evaluation Report.* Unpublished manuscript.

Krisberg, B., Schwartz, I., Fishman, G., Eisikovits, Z., Guttman, E., & Joe, K. (1987). The incarceration of minority youth. *Crime and Delinquency, 33*(2), 173–205.

Krisberg, B., & Vuong, L. (2009). *Rebuilding the infrastructure for at-risk youth.* Oakland, CA: National Council on Crime and Delinquency.

Land, K., Land, P., & Williams, J. (1990). Something that works in juvenile justice. *Evaluation Review, 14*(6), 574–606.

Langan, P., & Solari, R. (1993). *National judicial reporting program, 1990.* Washington, DC: U.S. Department of Justice, Office of Justice Programs, Bureau of Justice Statistics.

Latessa, E. (2008). *What works and what doesn't work in reducing recidivism.* Cincinnati, OH: University of Cincinnati.

Latessa, E., Turner, M., & Moon, M. (1998). *A statewide evaluation of the RECLAIM Ohio Initiative: Executive summary.* Columbus: Ohio Department of Youth Services.

Laub, J., & Sampson, R. (2003). *Shared beginnings: Delinquent boys to age 70.* Cambridge, MA; Harvard University Press.

Lawrence, C. (1987). The id, the ego and equal protection: Reckoning with unconscious racism. *Stanford Law Review, 39,* 317–388.

Le, T., Arifuku, I., Louie, C., & Krisberg, M. (2001). *Not invisible: Asian Pacific Islander juvenile arrests in San Francisco County.* Oakland, CA: National Council on Crime and Delinquency, Asian/Pacific Islander Youth Violence Prevention Center.

Le, T., Arifuku, I., Louie, C., Krisberg, M., & Tang, E. (2001). *Not invisible: Asian Pacific Islander juvenile arrests in Alameda County.* Oakland, CA: National Council on Crime and Delinquency, Asian/Pacific Islander Youth Violence Prevention Center.

Lee, L., & Zahn, G. (1998). Psychological status of children and youths. In L. C. Lee & N. Zane (Eds.), *Handbook of Asian American psychology.* Thousand Oaks, CA: Sage.

Lee, V., & Croninger, R. G. (1996). The social organization of safe high schools. In K. M. Borman, P. W. Cookson, Jr., & J. Z. Spade (Eds.), *Implementing educational reform: Sociological perspectives on educational policy* (pp. 359–392). Norwood, NJ: Ablex.

Leffert, N., Benson, P. L., Scales, P. C., Sharma, A. R., Drake, D. R., & Blyth, D. A. (1998). Developmental assets: Measurement and prediction of risk behaviors among adolescents. *Applied Developmental Science, 2*(4), 209–230.

Lejins, P. (1967). The field of prevention. In W. E. Amos & C. E. Wellford (Eds.), *Delinquency prevention: Theory and practice* (pp. 1–21). Englewood Cliffs, NJ: Prentice Hall.

Lemert, E. (1951). *Social pathology: A systematic approach to the theory of sociopathic behavior.* New York, NY: McGraw Hill.

Lemert, E., & Dill, F. (1978). *Offenders in the community: The probation subsidy in California.* Lexington, MA: Lexington Books.

Lemert, E., & Rosenberg, J. (1948). *The administration of justice to minority groups in Los Angeles County.* Berkeley: University of California Press.

Leonard, K., Pope, C., & Feyerherm, W. (Eds.). (1995). *Minorities in juvenile justice.* Thousand Oaks, CA: Sage.

Lerman, P. (1975). *Community treatment and social control: A critical analysis of juvenile correctional policy.* Chicago, IL: University of Chicago Press.

Lerner, G. (1973). *Black women in white America.* New York, NY: Random House Vintage.

Liazos, A. (1974). Class oppression and the juvenile justice system. *Insurgent Sociologist, 1,* 2–22.

Lindsey, B. B., & Borough, R. (1931). *The dangerous life.* New York, NY: Arno.

Lipsey, M. W. (1992). The effect of treatment on juvenile delinquents: Results from meta-analysis. In F. Loesel, D. Bender, & T. Bliesener (Eds.), *Psychology and law: International perspectives* (pp. 131–143). Berlin, NY: Walter de Gruyter.

Lipsey, M., Landenberger, N., & Wilson, S. (2007). *Effects of cognitive-behavioral programs for criminal offenders.* Nashville, TN: Center for Evaluation Research and Methodology, Vanderbilt Institute for Public Policy Studies.

Lipsey, M. W., & Wilson, D. B. (1998). Effective interventions with serious juvenile offenders: A synthesis of research. In R. Loeber & D. P. Farrington (Eds.), *Serious and violent juvenile offenders: Risk factors and successful interventions* (pp. 313–345). Thousand Oaks, CA: Sage.

Lipsey, M. W., & Wilson, D. B. (2001). *Practical meta-analysis*. Thousand Oaks, CA: Sage.

Loeber, R., & Farrington, D. P. (Eds.). (1998). *Serious and violent juvenile offenders: Risk factors and successful interventions*. Thousand Oaks, CA: Sage.

Loeber, R., & Farrington, D. P. (Eds.). (2001). *Child delinquents: Development, intervention, and service needs*. Thousand Oaks, CA: Sage.

Loeber, R., & Hay, D. F. (1997). Key issues in the development of aggression and violence from early childhood to early adulthood. *Annual Review of Psychology, 48*, 371–410.

Lombroso, C. (1911). *Crime: Its causes and remedies*. Boston, MA: Little, Brown.

Lubove, R. (1965). *The professional altruist: The emergence of social work as a career, 1880–1930*. Cambridge, MA: Harvard University Press.

Macallair, D. (2015). *After the doors were locked: A history of youth corrections in California and the origins of twenty-first century reform*. Lanthan, MD: Rowman & Littlefield.

MacKenzie, D. L. (2000). Evidence-based corrections: Identifying what works. *Crime & Delinquency, 46*, 457–471.

MacKenzie, D. L., Grover, A. R., Armstrong, G. S., & Mitchell, O. (2001). *National study comparing the environments of boot camps with traditional facilities for juvenile offender*. Washington, DC: U.S. Department of Justice, Office of Justice Programs, National Institute of Justice.

Maguire, K., & Pastore, A. (Eds.). (2000). *Sourcebook of criminal justice statistics*. Washington, DC: U.S. Department of Justice, Bureau of Justice Statistics.

Males, M. A., & Macallair, D. (2000). *The color of justice: An analysis of juvenile adult court transfers in California*. San Francisco, CA: Justice Policy Institute.

Mann, C. R. (1984). *Female crime and delinquency*. Tuscaloosa: University of Alabama Press.

Marans, S., Adnopoz, J., Berkman, M., Esserman, D., MacDonald, D., Nagler, S., . . . Wearing, M. (1995). *The police-mental health partnership: A community-based response to urban violence*. New Haven, CT: Yale University Press.

Markman, H., Renick, M., Floyd, F., Stanley, S., & Clements, M. (1993). Preventing marital distress through effective communication and conflict management: A 4- and 5-year follow-up. *Journal of Consulting and Clinical Psychology, 61*(1), 70–77.

Marris, P., & Rein, M. (1967). *Dilemmas of social reform: Poverty and community action in the United States*. New York, NY: Atherton.

Martinson, R. (1974). What works? Questions and answers about prison reform. *Public Interest, 35*, 22–54.

Mathias, R. A., DeMuro, P., & Allinson, R. (Eds.). (1984). *Violent juvenile offenders: An anthology*. San Francisco, CA: National Council on Crime and Delinquency.

Mathieson, T. (1965). *The defences of the weak*. London, England: Travistock.

Maton, K., Seifert, K., & Zapert, K. (1991). *Choice program evaluation: Preliminary findings*. Catonsville: University of Maryland, Department of Psychology.

Mattick, H., & Caplan, W. (1967). Stake animals, loud-talking and leadership in do-nothing and do-something situations. In M. Klein (Ed.), *Juvenile gangs in context* (pp. 121–135). Englewood Cliffs, NJ: Prentice Hall.

Maxson, C., Hennigan, K., & Sloane, D. (2006). It's getting crazy out there: Can a civil gang injunction change a community? *Criminology and Public Policy, 4*, 577–605.

Mayer, J. (1994). *Girls in the Maryland juvenile justice system: Findings of the female population taskforce*. Presentation to the Gender Specific Services Training, Minneapolis, MN.

McKelvey, B. (1972). *American prisons*. Montclair, NJ: Patterson Smith.

McKiever v. Pennsylvania, 403 U.S. 528 (1971).

Mead, G. (1934). *Mind, self and society*. Chicago, IL: University of Chicago Press.

Mead, G. (1961). The psychology of punitive justice. In T. Parsons (Ed.), *Theories of society* (pp. 878–894). Glencoe, IL: Free Press.

Melton, A. (1998). American Indians: Traditional and contemporary tribal justice. In C. Mann & S. Zatz (Eds.), *Images of color, images of crime: Readings* (pp. 58–71). Los Angeles, CA: Roxbury

Mendel, R. (2011). *No place for kids*. Baltimore MD: The Annie E. Casey Foundation.

Mendel, R. (2014). *Closing Massachusetts training schools: Reflections forty years later*. Baltimore, MD: The Annie E. Casey Foundation.

Mennel, R. (1973). *Thorns and thistles*. Hanover: The University of New Hampshire Press.

Merton, R. (1938). Social structure and anomie. *American Sociological Review, 3*, 672–82.

Miller, J. (1998). *Last one over the wall: The Massachusetts experiment in closing reform schools* (2nd ed.). Columbus: Ohio State University Press.

Miller, W. (1958). Lower class culture as a generating milieu of gang delinquency *Journal of Social Issues, 14,* 5–19.

Miller v. Alabama, 132 S. Ct. 2455 (2012).

Mintz, S. (2004). *Huck's raft: A history of American childhood.* Cambridge, MA: Harvard University Press.

Moffit, T. (1993). Adolescence-limited and life-course-persistent antisocial behavior: A developmental taxonomy. *Psychological Review, 100,* 674–701.

Moone, J. (1997). *Counting what counts: The census of juveniles in residential placement.* Washington, DC: U.S. Department of Justice, Office of Justice Programs, Office of Juvenile Justice and Delinquency Prevention.

Moore, J., & Hagedorn, J. (1988). Youth gangs: Milwaukee and Los Angeles compared. *Research and Opinion, 5*(1).

Moore, J., & Hagedorn, J. (2001). *Female gangs: A focus on research.* Washington, DC: U.S. Department of Justice, Office of Justice Programs, Office of Juvenile Justice and Delinquency Prevention.

Morris, M. (2002, June). Black girls on lockdown. *Essence, 33*(2), 186.

Moynihan, D. (1969). *Maximum feasible misunderstanding.* New York, NY: Free Press.

Murray, C., & Cox, L. (1979). *Beyond probation: Juvenile corrections and the chronic delinquent.* Beverly Hills, CA: Sage

Muscar, J. (2008). Advocating for the end of juvenile boot camps: Why the military model does not belong in juvenile justice. *UC Davis Journal of Juvenile Law and Policy, 12,* 1–50.

National Advisory Committee on Criminal Justice Standards and Goals. (1976). *Juvenile justice and delinquency prevention: Report of the task force on juvenile justice and delinquency prevention.* Washington, DC: Author.

National Commission on the Causes and Prevention of Violence. (1969). *To establish justice, to ensure domestic tranquility.* Washington, DC: Government Printing Office.

National Council on Crime and Delinquency. (1997). *Reducing crime in America.* San Francisco, CA: Author.

National Council on Crime and Delinquency. (2000). *Promising approaches for graduated sanctions.* Oakland, CA: Author.

National Council on Crime and Delinquency. (2001). *What works?: The R.Y.S.E. program, Alameda County Probation Department.* Oakland, CA: Author.

National Council on Crime and Delinquency. (2010). *An analysis of Asian and Pacific Islander youth in the California Youth Authority.* Oakland, CA: Author.

Nelson, L., & Lind, D. (2015). *The school to prison pipeline, explained.* Washington, DC: Justice Policy Institute.

Norman, S. (1972). *The youth service bureau: A key to delinquency prevention.* Hackensack, NJ: National Council on Crime and Delinquency.

Obama, B. (2016, January 25). Why we must rethink solitary confinement. *The Washington post.* Retrieved from https://www.washingtonpost.com/opinions/barack-obama-why-we-must-rethink-solitary-confinement/2016/01/25/29a361f2-c384-11e5-8965-0607e0e265ce_story.html?utm_term=.e7cd-1c3bb0bb

Obeidallah, D., & Earls, F. (1999, July). *Adolescent girls: The role of depression in the development of delinquency* [Research Preview]. Washington, DC: U.S. Department of Justice, Office of Justice Programs, National Institute of Justice.

O'Conner, S. (2001). *Orphan trains; The story of Charles Loring Brace and the children he saved and failed.* Boston, MA: Houghton Mifflin

Office of Juvenile Justice and Delinquency Prevention. (1998, December). *Guide for implementing the balanced and restorative justice model.* Washington, DC: U.S. Department of Justice.

Office of Juvenile Justice and Delinquency Prevention. (1999). *Juvenile arrests 1998.* Washington, DC: U.S. Department of Justice.

Office of Juvenile Justice and Delinquency Prevention. (2015). *Girls in the juvenile justice system: Policy guidance.* Washington, DC: U.S. Department of Justice.

Olds, D., Henderson, C., Chamberlin, R., & Tatelbaum, R. (1986). Preventing child abuse and neglect: A randomized trial of nurse home visitation. *Pediatrics, 78,* 65–78.

Olds, D., Hill, P., & Rumsey, E. (1998). *Prenatal and early childhood nurse home visitation*. Washington, DC: U.S. Department of Justice, Office of Justice Programs, Office of Juvenile Justice and Delinquency Prevention.

Olds, D., & Kitzman, H. (1993). Review of research on home visiting for pregnant women and parents of young children. *The Future of Children, 3*(3), 53–92.

Oliver, W., & Hairston, C. (2008). Intimate partner violence during the transition from prison to community: Perspectives of African American men. *Journal of Aggression, Maltreatment and Trauma, 16*, 258–276.

Ong, P., & Miller, D. (2002). *Economic needs of Asian Americans and Pacific Islanders in distressed areas: Establishing baseline information*. Washington, DC: U.S Department of Commerce, Research and National Technical Assistance, Economic Development Administration.

Oregon Social Learning Center. (2016). Various publications. Eugene: University of Oregon.

Osofsky, J. (2001). *Addressing youth victimization: Action plan update*. Washington, DC: U.S. Department of Justice, Office of Justice Programs, Office of Juvenile Justice and Delinquency Prevention, Coordinating Council on Juvenile Justice and Delinquency Prevention.

Palmer, T. (1971). California's community treatment program for delinquent adolescents. *Journal of Research in Crime and Delinquency, 8*(1), 74–82.

Palmer, T. (1992). *The re-emergence of correctional interventions*. Newbury Park, CA: Sage.

Palmer, T. (1996). Programmatic and nonprogrammatic aspects of successful intervention. In A. T. Harland (Ed.), *Choosing correctional options that work: Defining the demand and evaluating the supply* (pp. 131–182). Thousand Oaks, CA: Sage.

Parent, D., Lieter, V., Kennedy, S., Livens, L., Wentworth, D., & Wilcox, S. (1994). *Conditions of confinement: Juvenile detention and corrections facilities: Research report*. Washington, DC: U.S. Department of Justice, Office of Justice Programs, Office of Juvenile Justice and Delinquency Prevention.

Park, P., Burgess, E., & McKenzie, R. (1925). *The city*. Chicago: University of Chicago Press.

Parry, D. (1996). Race and community in juvenile justice decision making: Native Americans and the convergence of minority status and rural residence. *Social Pathology, 2*(2), 120–153.

Pasko, L., Okamoto, S., & Chesney-Lind, M. (2010). What about the girls: Gender, delinquency and juvenile justice in the 21st century. In D. W. Springer & A. W. Roberts (Eds.), *Juvenile Justice and Juvenile Delinquency*. Sudbury, England: Jones & Bartlett.

Patterson, G., DeBaryshe, B., & Ramsey, E. (1989). A developmental perspective on adolescent antisocial behavior. *American Psychologist, 44*, 329–335.

Pawasarat, J. (1991). *Identifying Milwaukee youth in critical need of intervention: Lessons from the past, measures for the future*. Milwaukee: University of Wisconsin, Employment and Training Institute.

The People v. Turner, 55 Illinois 280 (1870).

Perkins, C. (1994). *National corrections reporting program, 1992*. Washington, DC: U.S. Department of Justice, Office of Justice Programs, Office of Juvenile Justice and Delinquency Prevention.

Peters, M., Thomas, D., & Zamberlan, C. (1997). *Boot camps for juvenile offenders: Program summary*. Washington, DC: U.S. Department of Justice, Office of Justice Programs, Office of Juvenile Justice and Delinquency Prevention.

Peterson, D., & Truzzi, M. (Eds.). (1972). *Criminal life: Views from the inside*. Englewood Cliffs, NJ: Prentice Hall.

Pickett, R. (1969). *House of refuge: Origins of juvenile justice reform in New York, 1815–1857*. Syracuse, NY: Syracuse University Press.

Piliavin, I., & Briar, S. (1964). Police encounters with juveniles. *American Journal of Sociology, 70*(2), 206–214.

Platt, A. (1968). *The child savers: The invention of delinquency*. Chicago, IL: Chicago University Press.

Podkopacz, M., & Feld, B. (1996). The end of the line: An empirical study of judicial waiver. *Journal of Criminal Law and Criminology, 86*(2), 449–492.

Poe-Yamagata, E., & Butts, J. (1996). *Female offenders in the juvenile justice system*. Pittsburgh, PA: National Center for Juvenile Justice.

Poe-Yamagata, E., & Jones, M. (2000). *And justice for some*. Davis, CA: National Council on Crime and Delinquency.

Poe-Yamagata, E., Jones, M., Hartney, C., & Silva, F. (2007). *And justice for some. Differential treatment of youth of color in the justice system*. Oakland, CA: National Council on Crime and Delinquency.

Poe-Yamagata, E., Wordes, M., & Krisberg, B. (2002). *Young women and delinquency: Refining the research agenda* (A concept paper submitted to the MacArthur Foundation). Available from the National Council on Crime and Delinquency.

Pope, C., Lovell, R., & Hsia, H. (2002). *Disproportionate minority confinement: A review of the literature from 1989 through 2001*. Washington, DC: U.S. Department of Justice, Office of Justice Programs, Office of Juvenile Justice and Delinquency Prevention.

Pound, R. (1957). *Guide to juvenile court judges.* New York, NY: National Probation and Parole Association.

President's Advisory Commission on Asian Americans and Pacific Islanders. (2001). *A people looking forward: Action for access and partnerships in the 21st century.* Rockville, MD: White House Initiative on Asian Americans and Pacific Islanders.

President's Commission on Law Enforcement and the Administration of Justice. (1967). *The challenge of crime in a free society.* Washington, DC: Government Publishing Office.

Prison Rape Elimination Act of 2003, Pub. L. No. 108–79, 117 Stat. 972 (2003).

Prison Reform Litigation Act of 1995, 42 U.S. Code § 1997e (1995).

Rafter, N. (1997). *Creating born criminals.* Chicago: University of Illinois Press.

Raine, A. (2014). *The anatomy of violence: The biological roots of crime.* New York, NY: Vintage Books.

Reckless, W. (1962). A noncausal explanation: Containment theory. *Excerpta Criminologica, 2,* 131–134.

Rhine, E. (1996). Something works: Recent research on effective correctional programming. In American Correctional Association, *The state of corrections: Proceedings American Correctional Association annual conferences, 1995.* Lanham, MD: American Correctional Association.

Richette, L. A. (1969). *The throwaway children.* Philadelphia, PA: Lippincott.

Ridolfi, L., Washburn, M., & Guzman, F. (2016). *The prosecution of youth as adults: A county-level of analysis of prosecutor direct file in California and its disparate impact on youth of color.* Oakland, CA: W. Haywood Burns Institute, Center on Juvenile and Criminal Justice, and National Center for Youth Law.

Rieff, P. (1959). *Freud: The mind of the moralist.* Chicago, IL: University of Chicago Press.

Rieff, P. (1966). *The triumph of the therapeutic.* New York, NY: Harper & Row.

R.J. et al. v. Bishop, consent decree (2009).

Roper v. Simmons 543 US 551 (2005).

Rothman, D. (1971). *The discovery of the asylum.* Boston, MA: Little, Brown.

Rowe, D. (2001). *Biology and crime.* New York, NY: Oxford Press.

Rusche, G., & Kirchheimer, O. (1939). *Punishment and social structure.* New York, NY: Columbia University Press.

Ryan, W. (1971). *Blaming the victim.* New York, NY: Random House.

Sanchez-Jankowski, M. S. (1991). *Islands in the street: Gangs & American urban society.* Berkeley: University of California Press.

Sanders, W. B. (1970). *Juvenile offenders for a thousand years.* Chapel Hill: University of North Carolina.

Schall v. Martin, 467 U.S. 253 (1984).

Schur, E. M. (1973). *Radical nonintervention: Rethinking the delinquency problem.* Englewood Cliffs, NJ: Prentice Hall.

Schwartz, I. M., & Barton, W. H. (Eds.). (1994). *Reforming juvenile detention: No more hidden closets.* Columbus: Ohio State University Press.

Schwendinger, H., & Schwendinger, J. (1985). *Adolescent subcultures and delinquency.* New York, NY: Praeger.

Sechrest, D. (1970). *The community approach.* Berkeley: University of California, School of Criminology.

Sellin, T. (1938). *Culture conflict and crime.* New York, NY: Social Science Research Council.

Sellin, T. (1944). *Pioneering in penology.* Philadelphia: University of Pennsylvania Press.

Sellin, T. (2016). *Slavery and the penal system.* New York, NY: Elsevier.

Shaffer, A. (1971). The Cincinnati social unity experiment. *Social Service Review, 45,* 159–171.

Shakur, S. (1993). *Monster: The autobiography of an L.A. gang member.* New York, NY: Atlantic Monthly Press.

Shaw, C. R. (1930). *The jack-roller, a delinquent boy's own story* [Series title: Behavior research fund monographs]. Chicago, IL: The University of Chicago Press.

Shaw, C. (1938). *Brothers in crime* [Series title: Behavior research fund monographs]. Chicago, IL: University of Chicago Press.

Shaw, R., & McKay, L. (Eds.). (1942). *Juvenile delinquency and urban areas.* Chicago, IL: University of Chicago Press.

Sherman, L. (1990). Police crackdowns. *National Institute of Justice Reports, 219,* 2–6.

Sherman, L. W., Gottfredson, D., McKenzie, D. L., Eck, J., Reuter, P., & Bushway, S. (1997). *Preventing crime: What works, what doesn't, what's promising. Report to the U.S. Congress.* Washington, DC: U.S. Department of Justice, Office of Justice Programs, National Institute of Justice.

Sickmund, S., & Puzzanchera, C. (Eds.) (2014). *Juvenile offenders and victims: 2014 national report.* Pittsburgh, PA: National Center for Juvenile Justice.

Singer, S., & McDowall, D. (1988). Criminalizing delinquency: The deterrent effects of the New York juvenile offender law. *Law and Society Review, 22,* 521–535.

Skogan, W., Hartnett, S., Bump, N., & Dubois, J. (2009). *The Evaluation of cease-fire Chicago.* Washington, DC: National Institute of Justice.

Smith, B. (1998). Children in custody: 20-year trends in juvenile detention, correctional, and shelter facilities. *Crime and Delinquency, 44*(4), 526–543.

Snyder, H. (1999). *Juvenile arrests 1998.* Washington, DC: U.S. Department of Justice, Office of Justice Programs, Office of Juvenile Justice and Delinquency Prevention.

Snyder, H., Finnegan, T., Stahl, A., & Poole, R. (1999). *Easy access to juvenile court statistics: 1988–1997* [data presentation and analysis package]. Pittsburgh, PA: National Center for Juvenile Justice (producer); Washington, DC: Office of Juvenile Justice and Delinquency Prevention (distributor).

Snyder, H., & Sickmund, M. (1999). *Juvenile offenders and victims: 1999 national report.* Washington, DC: U.S. Department of Justice, Office of Justice Programs, Office of Juvenile Justice and Delinquency Prevention.

Sok, C. W. (2001). Southeast Asian Americans and education. *The Bridge, 19,* 1–2.

Spergel, I., & Grossman, S. (1996). *Evaluation of a gang violence reduction project: A comprehensive and integrated approach.* Chicago, IL: University of Chicago, School of Social Service Administration.

Spradley, J. (1999). *You owe yourself a drunk: An ethnography of urban nomads.* Long Grove, IL: Waveland Press.

Stahl, A. (1998). *Delinquency cases in juvenile courts: Fact sheet #79, 1995.* Washington, DC: U.S. Department of Justice, Office of Justice Programs, Office of Juvenile Justice and Delinquency Prevention.

Stamp, K. (1956). *The peculiar institution.* New York, NY: Random House.

State of California, Department of the Youth Authority. (n.d.). *A comparison of first commitment characteristics 1990–2001.* Sacramento, CA: Administrative Services Branch, Research Division, Ward Information, and Parole Research Bureau.

State v. Ray, 63 New Hampshire 405 (1886).

Steiner, H., Garcia, I., & Mathews, Z. (1997). Post-traumatic stress in incarcerated juvenile delinquents. *Journal of the American Academy of Child and Adolescent Psychiatry, 36,* 357–365

Steinhart, D. (1988). California legislature ends the jailing of children: The story of a policy reversal. *Crime and Delinquency, 32*(2), 169–189.

Steinhart, D. (1994, August). Assessing the need for secure detention: A planning approach. *NCCD Focus.* San Francisco: The National Council on Crime and Delinquency.

Strehike, S. (2015, October 30). Spring Valley students walk out in support of the officer fired for beating up a girl. *Teen Vogue.* Retrieved from http://www.teenvogue.com/story/spring-valley-high-school-student-dragged-school-officer

Streib, V. (1987). *Death penalty for juveniles.* Indianapolis: Indiana University Press.

Strom, K. (2000). *Profile of state prisoners under age 18.* Washington, DC: U.S. Department of Justice, Office of Justice Programs, Bureau of Justice Statistics.

Sue, S., & Kitano, H. (Issue Eds.). (1973). Asian Americans: A success story? *Journal of Social Issues, 29*(2).

Sutherland, E. (1947). *Principles of criminology.* Philadelphia, PA: Lippincott.

Sykes, G., & Matza, D. (1957). Techniques of neutralization: A theory of delinquency. *American Sociological Review, 22,* 664–670.

Takagi, P. (1974). The correctional system. *Crime and Social Justice, 2,* 82–87.

Takagi, R. (1989). *Strangers from a different shore: A history of Asian Americans.* Boston, MA: Little, Brown.

Taylor, I., Walton, P., & Young, J. (1975). *The new criminology: For a social theory of deviance.* London, England: Routledge.

Thomas, P. (1967). *Down these mean streets.* New York, NY: Vintage.

Thompson v. Oklahoma, 487 U.S. 815 (1987).

Thornberry, T. (1994). *Violent families and youth violence: Fact sheet #21.* Washington, DC: U.S. Department of Justice, Office of Justice Programs, Office of Juvenile Justice and Delinquency Prevention.

Thornberry, T., Huizinga, D., & Loeber, R. (1995). The prevention of serious delinquency and violence: Implications from the program of research on the causes and correlates of delinquency. In J. C. Howell, B. Krisberg, D. Hawkins, & J. Wilson (Eds.), *Serious, violent, & chronic juvenile offenders: A sourcebook* (pp. 213–237). Thousand Oaks, CA: Sage.

Thrasher, F. M. (1927). *The gang.* Chicago, IL: University of Chicago Press.

Tollett, T. (1987). *A comparative study of Florida delinquency commitment programs.* Tallahassee: Florida Department of Health and Rehabilitative Services.

Torbet, P., Gable, R., Hurst, I., IV, Montgomery, L., Szymanski, L., & Thomas, D. (1996). *State responses to serious and violent juvenile crime.* Washington, DC: U.S. Department of Justice, Office of Justice Programs, Office of Juvenile Justice and Delinquency Prevention.

Turk, A. (1969). *Criminality and legal order.* Chicago, IL: Rand McNally

U.S. Bureau of Justice Statistics. (2013). *National survey of youth in custody, 2013.* Washington, DC: Author.

U.S. Department of Health and Human Services, Administration of Children, Youth and Families. (1999). *Child maltreatment 1997: Reports from the states to the National Child Abuse and Neglect Data System.* Washington, DC: U.S. Government Printing Office.

U.S. Department of Health, Education, and Welfare. (1973). *National study of youth service bureaus—Final report.* Washington, DC: California Youth Authority.

U.S. Department of Justice. (2016). *Report and recommendations concerning the use of restricted housing.* Washington, DC: Author.

Villarruel, F. A., & Walker, N. E. (with Minefee, P., Rivera-Vazquez, O., Peterson, S., & Perry, K.). (2002). *¿Dónde está la justicia? A call to action on behalf of Latino and Latina youth in the U.S. justice system.* Lansing: Michigan State University Institute for Children, Youth, & Families.

Vold, G. (1958). *Theoretical criminology.* Newark: University of Delaware Press.

Vuong, L., & Silva, F. (2008). *Evaluating federal crime bills.* Oakland, CA: National Council on Crime and Delinquency.

Walker, J., Cardarelli, A., & Billingsley, D. (1976). *The theory and practice of delinquency prevention in the United States: Review, synthesis and assessment.* Columbus: Ohio State University, Center for Vocational Education.

Walker, M. (1969). *Jubilee.* New York, NY: Bantam.

Ward, G. (2012). *The black child-savers: Racial democracy and juvenile justice.* Chicago, IL: University of Chicago Press.

Weiner, N. (1996). *The priority prosecution of the serious habitual juvenile offender: Roadblocks to early warning, early intervention, and maximum effectiveness—the Philadelphia study.* Philadelphia: University of Pennsylvania, School of Social Work, Center for the Study of Youth Policy.

Weinstein, J. (1968). *The corporate ideal in the liberal state.* Boston, MA: Beacon Press.

Weiss, F., Nicholson, H., & Cretella, M. (1996). *Prevention and parity: Girls in juvenile justice.* New York, NY: Girls Incorporated Nation Resource Center.

Weissman, J. (1969). *Community development in the mobilization for youth.* New York, NY: Association Press.

White, J. (1985). *The comparative dispositions study: A report.* Washington, DC: U.S. Department of Justice, Office of Justice Programs, Office of Juvenile Justice and Delinquency Prevention.

Widom, C. (1989). Does violence beget violence? A critical examination of the literature. *Psychological Bulletin, 106,* 3–28.

Widom, C. (1992). *The cycle of violence.* Washington, DC: U.S. Department of Justice, Office of Justice Programs, National Institute of Justice.

Widom, C. S. (1995). *Victims of childhood sexual abuse: Later criminal consequences.* Washington, DC: U.S. Department of Justice, Office of Justice Programs, National Institute of Justice.

Wiebush, R. G. (1993). Juvenile intensive supervision: The impact on felony offenders diverted from institutional placement. *Crime and Delinquency, 39*(1), 68–89.

Wiebush, R. G., Baird, C., Krisberg, B., & Onek, D. (1995). Risk assessment and classification for serious, violent, and chronic juvenile offenders. In J. C. Howell, B. Krisberg, J. D. Hawkins, & J. J. Wilson (Eds.), *Serious, violent, & chronic juvenile offenders: A sourcebook* (pp. 171–212). Thousand Oaks, CA: Sage.

Wilhelm, S. (1970). *Who needs the Negro?* Cambridge, MA: Schenkman.

Williams, W. (1973). *The contours of American history.* New York, NY: New Viewpoints.

Wilson, D., Mackenzie, D., & Mitchell, D., (2008). *The effect of correctional boot camps on criminal offending.* Philadelphia, PA: The Campbell Collaboration.

Wilson, J. (1968). *Varieties of police behavior: The management of law and order in eight communities.* Cambridge, MA: Harvard University Press.

Wilson, J. (1983). *Thinking about crime* (Rev. ed.). New York, NY: Basic.

Wilson, J. (1995). Crime and public policy. In J. Q. Wilson & J. Petersilia (Eds.), *Crime* (pp. 489–507). San Francisco, CA: Institute for Contemporary Studies Press.

Wilson, J., & Herrnstein, R. (1985). *Crime and human nature.* New York, NY: Simon & Schuster.

Wilson, J., & Howell, J. (1993). *A comprehensive strategy for serious, violent, and chronic juvenile offenders.* Washington, DC: U.S. Department of Justice, Office of Justice Programs, Office of Juvenile Justice and Delinquency Prevention.

Wines, E. (1880). *The state of prisons and child-saving institutions in the civilized world.* Cambridge, England: J. S. Wilson & Son.

Wirth, L. (1928). *The ghetto.* Chicago, IL: University of Chicago Press.

Witmer, J., & Tufts, E. (1954). *The effectiveness of delinquency prevention programs.* Washington, DC: Government Printing Office.

Wolfgang, M., Figlio, R., & Sellin, T. (1972). *Delinquency in a birth cohort.* Chicago, IL: University of Chicago Press.

Wolfgang, M., Thornberry, T., & Figlio, R. (1987). *From boy to man, from delinquency to crime.* Chicago, IL: University of Chicago Press.

Wordes, M., & Nuñez, M. (2002). *Our vulnerable teenagers: Their victimization, its consequences, and directions for prevention and intervention.* Oakland, CA: National Council on Crime and Delinquency.

Wright, W. E., & Dixon, M. C. (1977). Community prevention and treatment of juvenile delinquency: A review of evaluation studies. *Journal of Research in Crime and Delinquency, 14*(1), 35–67.

Yetman, N. R. (1970). *Voices from slavery.* New York, NY: Holt, Rinehart and Winston.

Yoshikawa, H. (1994). Prevention as cumulative protection: Effects of early family support and education on chronic delinquency and its risks. *Psychological Bulletin, 115,* 1–27.

Yoshikawa, H. (1995). Long-term effects of early childhood programs on social outcomes and delinquency. *The Future of Children, 5*(3), 51–75.

Zahn, M., Hawkins, S., Chiancone, J., & Witworth, A. (2008). *The girls study group: Charting the way for delinquency prevention for girls.* Washington, DC: Office of Justice Programs.

Zimring, F. (2005). *American juvenile justice.* New York, NY: Oxford University Press.

Zimring, F. (2016). *American youth violence.* New York, NY: Oxford University Press.

Zimring, F. (in press). Minimizing harm from minority disproportion in American juvenile justice. In D. Hawkins & K. Kempf-Leonard (Eds.), *Our children, their children: Confronting racial and ethnic differences in American juvenile justice.* Chicago, IL: University of Chicago Press.

Zimring, F., & Hawkins, G. (1973). *Deterrence: The legal threat in crime control.* Chicago, IL: University of Chicago Press.

Zorbaugh, H. (1929). *The gold coast and the slum.* Chicago, IL: University of Chicago Press.

INDEX

ABOUT THE AUTHOR

Barry Krisberg, PhD, was the president of the National Council on Crime and Delinquency from 1983 until 2010. Between 2010 and 2015, he was at the University of California, Berkeley, School of Law. He is known nationally for his research and expertise on juvenile justice issues and is called upon as a resource for professionals and the media. He is currently a visiting fellow at the Institute for the Study of Societal Issues at the University of California, Berkeley.

He received his master's degree in criminology and a doctorate in sociology, both from the University of Pennsylvania.

He has held several educational posts. He was a faculty member at the University of California at Berkeley; he was also an adjunct professor with the Hubert Humphrey Institute of Public Affairs at the University of Minnesota. He was a clinical professor of psychiatry at the University of Hawaii and a visiting lecturer at the University of California at Berkeley.

He was appointed by the legislature to serve on the California Blue Ribbon Commission on Inmate Population Management. He was a past president and fellow of the Western Society of Criminology and was the chair of the California Attorney General's Research Advisory Committee. In 1993, he was the recipient of the August Vollmer Award, the American Society of Criminology's most prestigious award. The Jessie Ball duPont Fund named him the 1999 Grantee of the Year for his outstanding commitment and expertise in the area of juvenile justice and delinquency prevention.

He has several books and articles to his credit, including *Crime and Privilege; The Children of Ishmael: Critical Perspectives on Juvenile Justice* with James Austin, PhD; *Juvenile Justice: Improving the Quality of Care; Reinventing Juvenile Justice* with James Austin, PhD; and *American Corrections Concepts and Controversies* with Susan Marchionna and Christopher Hartney.

He is frequently called upon by private foundations to assist with the design, implementation, and evaluation of programs—both their own and those of the public sector. For example, most recently, he was the chair on an expert panel reviewing the conditions and policies of the California Youth Authority. The Annie E. Casey Foundation engaged him to assess the effectiveness of their Juvenile Detention Alternatives Initiative, the single-largest private program for juvenile justice system reform in the country. The Walter Johnson Foundation asked that he lead a Blue Ribbon Commission on California's policy of out-of-state placement of delinquent youth. The California Endowment and the California Wellness Foundation supported his research on a number of juvenile programs. The Jesse Ball duPont Fund asked Dr. Krisberg to evaluate a number of programs in Florida. The Baptist Community Ministries has sought his expertise on assessing the needs of troubled youth in the greater New Orleans area, and the Robert Wood Johnson Foundation asked that he prepare a white paper on delinquency and substance abuse.

He regularly appears as an expert on national network news shows about criminal justice issues and is frequently consulted by the leading print media.